AQA Mathematics

for GCSE

Exclusively endorsed and approved by AQA

Series Editor
Paul Metcalf
Series Advisor
David Hodgson
Lead Author
Steven Lomax

June Haighton
Anne Haworth

Margaret Thornton
Mark Willis

HIGHER
Linear 2

Published in 2006 by:
Nelson Thornes Ltd
Delta Place
27 Bath Road
CHELTENHAM
GL53 7TH
United Kingdom

07 08 09 10 / 10 9 8 7 6 5 4 3

A catalogue record for this book is available from the British Library.

978-0-7487-9753-0

Cover photograph: Seals by Stephen Frink/Digital Vision LU (NT)
Illustrations by Roger Penwill
Page make-up by MCS Publishing Services Ltd, Salisbury, Wiltshire

Printed in Great Britain by Scotprint

Acknowledgements

The authors and publishers wish to thank the following for their contribution:
David Bowles for providing the Assess questions
David Hodgson for reviewing draft manuscripts

Thank you to the following schools:
Little Heath School, Reading
The Kingswinford School, Dudley
Thorne Grammar School, Doncaster

The publishers thank the following for permission to reproduce copyright material:

Explore photos
Diver – Corel 55 (NT); Astronaut – Digital Vision 6 (NT);
Mountain climber – Digital Vision XA (NT); Desert explorer – Martin Harvey/Alamy.

Eurostar – Imagin London (NT); Planets – Digital Vision 9 (NT); Arab dhow – Corel 58 (NT);
Jogger – Photodisc 51(NT); Motorway jam – Digital Vision 15 (NT); Modern train – Corel 783 (NT);
Factory worker – Digital Stock 7 (NT); US Dollar bill – Photodisc 68 (NT);
Albert Einstein – Illustrated London News V2 (NT); Church – Corel 750 (NT);
Students sitting exam – Image 100 37 (NT); Cricketer – Corel 449 (NT); Volleyball – Corel 550 (NT);
Statue of Liberty – Corel 687 (NT); Supermarket – Stockbyte 9 (NT);
Lifeguard – Rubberball 68 (NT); Ferry – Corel 707 (NT); Louvre – Corel 437 (NT);
Stained glass – Corel 153 (NT); Coins – Alex Homer; MP3 player – Alex Homer;
Rabbit – Photodisc 50 (NT); Helicopter – Corel 415 (NT); Car engine – Alex Homer;
E.coli bacteria – NIBSC/Science Photo Library; Arsenic – Klaus Guldbrandsen/Science Photo Library

The publishers have made every effort to contact copyright holders but apologise if any have been overlooked.

Contents

Introduction

This book has been written by teachers and examiners who not only want you to get the best grade you can in your GCSE exam but also to enjoy maths.

Each chapter has the following stages:

OBJECTIVES

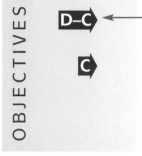 D–C

C

The objectives at the start of the chapter give you an idea of what you need to do to get each grade. Remember that the examiners expect you to perform well at the lower grade questions on the exam paper in order to get the higher grades. So, even if you are aiming for an A* grade you will still need to do well on the D grade questions on the exam paper.

Learn 1

Key information and examples to show you how to do each topic. There are several Learn sections in each chapter.

Apply 1

Questions that allow you to practise what you have just learned.

Means that these questions should be attempted with a calculator.

Means that these questions are practice for the non-calculator paper in the exam and should be attempted without a calculator.

Get Real!

These questions show how the maths in this topic can be used to solve real-life problems.

<u>1</u>

Underlined questions are harder questions.

Explore

Open-ended questions to extend what you have just learned. These are good practice for your coursework task.

ASSESS

End of chapter questions written by an examiner.

Some chapters feature additional questions taken from real past papers to further your understanding.

1 Properties of circles

OBJECTIVES

B Examiners would normally expect students who get a B grade to be able to:

Use the angle and tangent/chord properties of a circle

A Examiners would normally expect students who get an A grade also to be able to:

Prove the angle and tangent/chord properties of a circle

Use and prove the alternate segment theorem

What you should already know ...

- Recall the definition of a circle and the meaning of terms, including centre, radius, chord, diameter, circumference, tangent, arc, sector and segment

- Understand basic proofs such as the angle sum of a triangle is 180°, the exterior angle of a triangle is equal to the sum of the interior opposite angles

- Understand angle properties of parallel lines including corresponding angles and alternate angles

VOCABULARY

Arc (of a circle) – part of the circumference of a circle; a minor arc is less than half the circumference and a major arc is greater than half the circumference

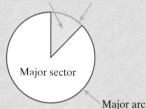

Minor sector   Minor arc

Major sector

Major arc

Subtend – when the end points of an arc are joined to a point on the circumference of a circle the angle formed is subtended by the arc

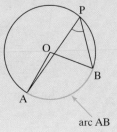

arc AB subtends ∠AOB at the centre

arc AB subtends ∠APB at the circumference

arc AB

Circumference – the perimeter of a circle

Cyclic quadrilateral – a quadrilateral whose vertices lie on the circumference of a circle

Supplementary – two angles are supplementary if their sum is 180°

Chord – a straight line joining two points on the circumference of a circle

Tangent (to a circle) – a straight line that touches the circle at only one point

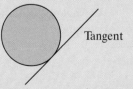

Tangent

Diameter – a chord passing through the centre of a circle; the diameter is twice the length of the radius

Radius – the distance from the centre of a circle to any point on the circumference

1

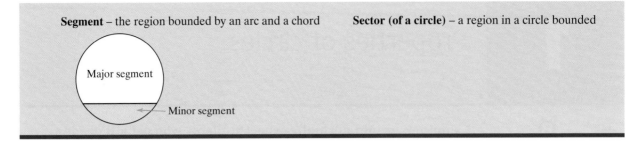

Segment – the region bounded by an arc and a chord   **Sector (of a circle)** – a region in a circle bounded

Major segment

Minor segment

Learn 1   Angle properties of a circle

The angle properties of a circle are:

1 The angle subtended by an arc (or chord) at the centre of a circle is twice the angle subtended at any point on the circumference.

Arc PQ subtends the angle POQ at the centre of the circle.

Arc PQ subtends angle PRQ at the circumference.

$\angle POQ = 2 \times \angle PRQ$

2 The angle (subtended) in a semicircle is a right angle.

The semicircular arc PQ subtends angle POQ at the centre of the circle and angle PRQ at the circumference.

$\angle POQ = 2 \times \angle PRQ$
$\angle POQ = 180°$ POQ is a straight line
so $\angle PRQ = 90°$

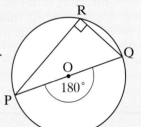

3 The opposite angles of a cyclic quadrilateral are supplementary (their sum is 180°).

$e + g = 180°$
$f + h = 180°$

4 Angles subtended by the same arc (or chord) are equal.   The angle at D and the angle at C are subtended by the same arc AB

Therefore these angles are the same:
Angle ADB = Angle ACB

Also in this diagram, the angle at A and the angle at B are subtended by the same arc DC
Angle DAC = Angle DBC

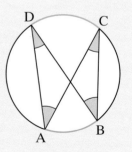

Not drawn accurately

2

Examples:

a The chord AB subtends an angle of 100° at the centre of the circle.
Calculate the angle ADB and the angle ACB giving reasons for your answers.

> The word 'calculate' in this type of question means that you must work out the angles, **not** measure them

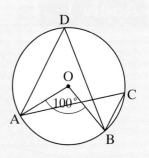

Not drawn accurately

Introduction
The chord AB subtends an angle of 100° at the centre of the circle.

> It is good practice to start your calculation by stating what you already know

Calculation
∠ADB = 50° (angle at centre = 2 × angle at circumference)

> These are the reasons and should be laid out carefully

∠ACB = 50° (angles subtended by the same arc AB are equal)

> Alternatively you could use the fact that:
> angle at the centre = 2 × angle at circumference

b AD is a diameter of the circle, centre O.
Angle ADC = 50° and angle ACB = 20°.
Calculate the angle DCA and the angle DAB giving reasons for your answers.

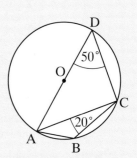

Introduction
AD is a diameter of the circle, centre O.

∠ADC = 50° and ∠ACB = 20°   Start by stating what you already know

Calculation

Not drawn accurately

∠DCA = 90°   (angle (subtended) in a semicircle is a right angle)

∠DCB = ∠DCA + ∠ACB
= 90° + 20°
= 110°

∠DAB = 180° − ∠DCB (opposite angles of a cyclic quadrilateral add up to 180°)
= 180° − 110°   Check that your answers are sensible
= 70°

3

Apply 1

1 Calculate the marked angle in each diagram, giving reasons for your answers. The centre of the circle is marked O.

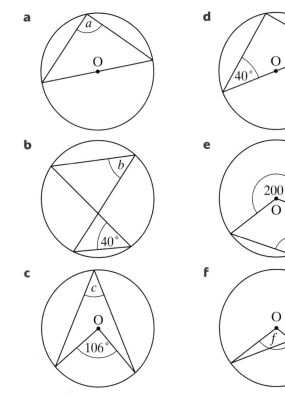

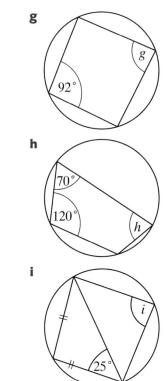

2 Are these statements true or false?

a Angles subtended by the same arc of a circle at points on the circumference are equal.

b The angle subtended by a chord at the centre of a circle is twice the angle subtended at the circumference.

c The opposite pairs of angles of a quadrilateral are supplementary.

d The angle subtended at the circumference by the diameter of a circle is a right angle.

e Adjacent pairs of angles in a cyclic quadrilateral are complementary (they add up to 90°).

f The angle subtended by a chord at the centre of a circle is half the angle subtended by the same chord at the circumference.

3 Write down all the angles that are the same size as angle ACE.

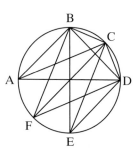

4 Adnan says that a parallelogram cannot be a cyclic quadrilateral.
Is he correct? Give a reason for your answer.

5 Sheila thinks angle ADC is 65° because the angle at the centre is twice
that subtended at the circumference.
Is Sheila correct?
Give a reason for your answer.

6 Bill thinks angle ABC is 50° because opposite angles of a cyclic
quadrilateral are supplementary.
Is Bill correct?
Give a reason for your answer.

7 The lines PR and QS pass through the centre of the circle at O.
Angle ORQ = 52°.
Calculate angles QSP and QOR.
Remember to show all your working.

8 LN is a diameter of the circle. Angle NLM = x and angle MNL = $2x$.
Work out the value of x.
What is the size of angle MNL?

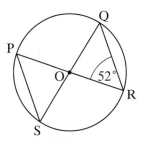

9 A triangle is drawn so that the three vertices lie on a circle with centre O.
Two of the angles of the triangle are 32° and 58°.
Jenny says that one of the sides of the triangle must pass through O.
Is she correct? Give a reason for your answer.

10 A quadrilateral is drawn inside a circle, centre O.
Angle OPQ = 65° and angle POR = 150°.
Calculate the other two angles of the quadrilateral.
Give reasons for your answers.

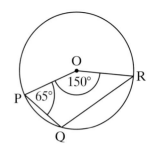

11 Prove the following circle properties.

    a The angles subtended by the same arc (or chord) are equal.

    b The angle subtended by an arc at the centre of a circle is twice the angle subtended at any point on the circumference.

    c The angle subtended at the circumference by a semicircle is a right angle.

    d Opposite angles of a cyclic quadrilateral are supplementary.

Explore

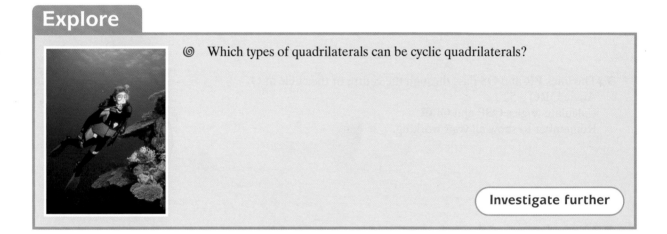

    ◎   Which types of quadrilaterals can be cyclic quadrilaterals?

Investigate further

Learn 2   Tangents and chords

The tangent/chord properties of a circle are:

1 The tangent at any point on a circle is perpendicular to the radius at that point.

2 Tangents from an external point are equal in length.

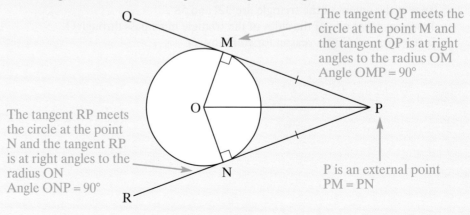

The tangent QP meets the circle at the point M and the tangent QP is at right angles to the radius OM
Angle OMP = 90°

The tangent RP meets the circle at the point N and the tangent RP is at right angles to the radius ON
Angle ONP = 90°

P is an external point
PM = PN

3 The perpendicular from the centre of a circle to a chord bisects the chord.

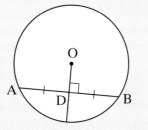

OD bisects the chord AB

4 Alternate segment theorem.

The alternate segment theorem says that the angle between tangent and chord is equal to the angle in the alternate segment.

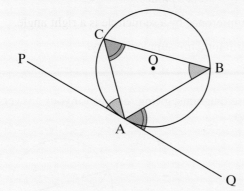

PQ is a tangent. It touches the circle at A.
∠PAC is an angle between the tangent and a chord AC.

The chord AC divides the circle into two segments.
∠PAC is (mostly) in one segment.
∠CBA is in the other (alternate) segment.

∠PAC = ∠CBA

Similarly, the angle between the tangent and the chord AB is equal to the angle subtended by the chord AB.

∠QAB = ∠ACB

Example:

PQ is a tangent to the circle, centre O. Angle CAP = 60°.
Calculate the angles ABC, AOC and OAC.

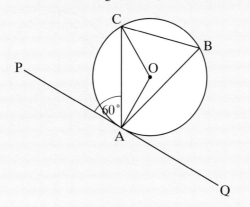

Not drawn accurately

Introduction
PQ is a tangent to the circle, centre O.

It is good practice to start by stating what you already know

Calculation
∠ABC = 60°   Alternate segment theorem

The order of the questions suggests that angle ABC is the easiest angle to find

∠AOC = 120°   Angle at centre = 2 × angle at circumference

∠OAC = 90° − ∠CAP   Tangent is perpendicular to the radius
      = 90° − 60°   Alternatively you could use the fact that angle OAC
      = 30°   is the base angle of the isosceles triangle OAC

Check that your answers are sensible

Apply 2

1 Calculate the marked angles in these diagrams, giving reasons for your answers.
   The centre of the circle is marked O.

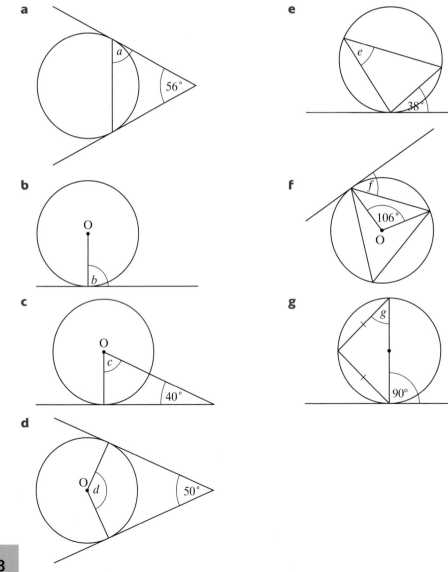

a

b

c

d

e

f

g

2 Decide whether these statements are true or false.

   a The perpendicular bisector of a chord passes through the centre of the circle.

   b The angle between the tangent and the chord is the same as the angle in the alternate segment.

   c A tangent is a straight line which touches a circle at one or more points on the circumference.

   d The tangent at any point on a circle is perpendicular to the radius at that point.

   e Tangents from an external point to the circumference of a circle are equal in length.

3 PR is a tangent to the circle, centre O. The tangent touches the circle at Q.
Angle QOS = 125°. Calculate the angle QRO.
Remember to show all your working.

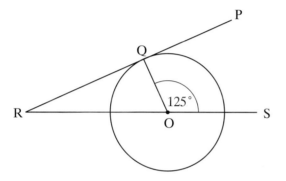

4 PQ is a tangent to the circle, centre O. Angle CAP = 55°.
Calculate the angles ABC, AOC and OAC.
Remember to show all your working.

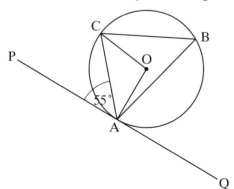

5 Tangents SP and SR meet the circle at P and R respectively.
SOQ is a straight line passing through the centre of the circle O. Angle PSO = 22°.
Calculate:

   a angle RSO

   b angle POR

   c angle PQR.

Give reasons for your answers.

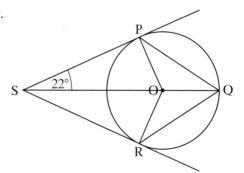

9

6 Assad says that angle CAP = angle CBA by the alternate segment theorem. Brian says that angle QAB = angle ACB by the alternate segment theorem.
Who is correct?
Give a reason for your answer.

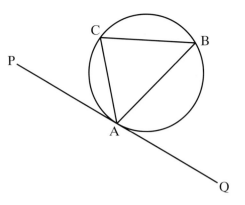

7 XY is a tangent to the circle, touching at the point A.
AB = BC and CD = DA. Angle XAB = 48°.
Calculate:

a angle BCA

b angle ABC

c angle DAC.

Remember to show all your working.

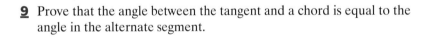

8 Tangents PR and PT meet the circle at Q and S respectively, WQ = WS. Show that angle RQW = angle TSW.

9 Prove that the angle between the tangent and a chord is equal to the angle in the alternate segment.

Explore

You can use the fact that the perpendicular bisector of a chord passes through the centre of a circle to find the centre of a circle.

To construct the centre of a given circle:

◎ pick three points on the circle A, B and C

◎ join AB and AC

◎ find the perpendicular bisectors of AB and AC

◎ mark the intersection, O, of these two bisectors

◎ O is the centre of the given circle

Investigate further

Properties of circles

The following exercise tests your understanding of this chapter, with the questions appearing in order of increasing difficulty.

In each of these diagrams find the marked angles.

In the diagrams, O is always the centre of the circle.

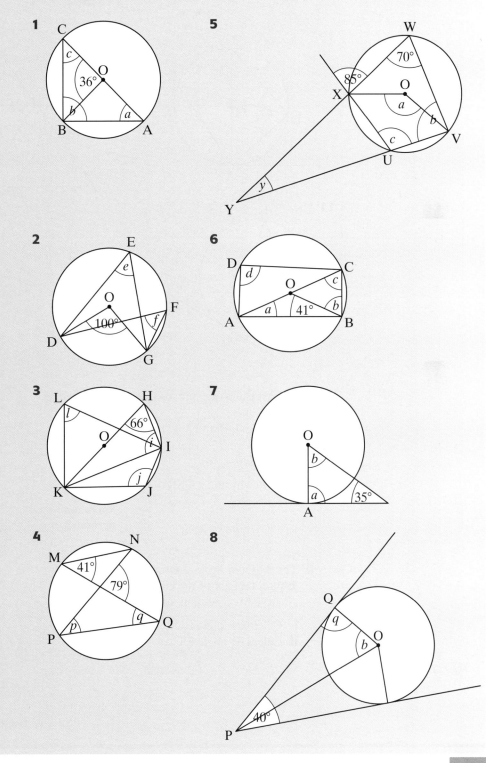

1

2

3

4

5

6

7

8

9

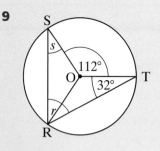

12

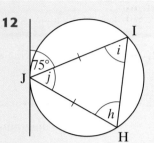

10

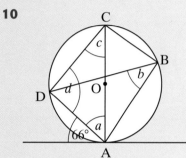

13

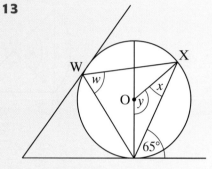

11

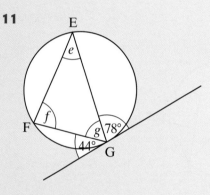

Try a real past exam question to test your knowledge:

14 a Points P, Q, R and S lie on a circle.
PQ = QR.
Angle PQR = 116°.

Explain why angle QSR = 32°.

Not drawn accurately

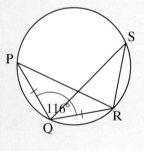

b The diagram shows a circle, centre O.
TA is a tangent to the circle at A.
Angle BAC = 58° and angle BAT = 74°.

i Calculate angle BOC.
ii Calculate angle OCA.

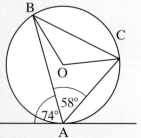

Spec A, Higher Paper 1, June 04

OBJECTIVES

C ▷

Examiners would normally expect students who get a C grade to be able to:

Form and solve equations such as $x^3 + x = 12$ using trial and improvement methods

What you should already know ...

■ Substitute into algebraic expressions

■ Use the bracket and power buttons on your calculator

VOCABULARY

Trial and improvement – a method for solving algebraic equations by making an informed guess then refining this to get closer and closer to the solution

Decimal places – the digits to the right of a decimal point in a number, for example, in the number 23.657, the number 6 is the first decimal place (worth $\frac{6}{10}$), the number 5 is the second decimal place (worth $\frac{5}{100}$), and 7 is the third decimal place (worth $\frac{7}{1000}$); the number 23.657 has 3 decimal places

Square number – a square number is the outcome when a whole number is multiplied by itself; square numbers are 1, 4, 9, 16, 25, ...

Square root – the square root of a number such as 16 is a number whose outcome is 16 when multiplied by itself

Cube number – a cube number is the outcome when a whole number is multiplied by itself then multiplied by itself again; cube numbers are 1, 8, 27, 64, 125, ...

Cube root – the cube root of a number such as 125 is a number whose outcome is 125 when multiplied by itself then multiplied by itself again

Learn 1 Trial and improvement

Examples:

a Use the trial and improvement method to solve $x^3 - 2x + 1 = -40$
Give your answer to two decimal places.

Laying out the information in tabular form:

Start with a good guess

Trial value of x	$x^3 - 2x + 1$	Comment
-3	$(-3)^3 - (2 \times -3) + 1 = -20$	Too high
-4	$(-4)^3 - (2 \times -4) + 1 = -55$	Too low
Now you know that the value lies between -3 and -4 Try -3.5		
-3.5	$(-3.5)^3 - (2 \times -3.5) + 1 = -34.875$	Too high
-3.6	$(-3.6)^3 - (2 \times -3.6) + 1 = -38.456$	Too high
-3.7	$(-3.7)^3 - (2 \times -3.7) + 1 = -42.253$	Too low
Now you know that the value lies between -3.6 and -3.7 Try -3.65		
-3.65	$(-3.65)^3 - (2 \times -3.65) + 1 = -40.327125$	Too low
-3.64	$(-3.64)^3 - (2 \times -3.64) + 1 = -39.948544$	Too high
Now you know that the value lies between -3.64 and -3.65 But you need to give the value to two decimal places Try -3.645		
-3.645	$(-3.645)^3 - (2 \times -3.645) + 1 = -40.13756113$	Too low
Now you know that the value lies between -3.645 and -3.65 But both values are -3.65 to two decimal places		

The required answer is $x = -3.65$ to two decimal places.

b The area of a rectangle is 63 cm^2.
One side is 4 cm longer than the other side as shown.

$(x+4)$

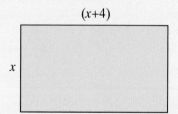

x

Use the trial and improvement method to find the value of x.

Give your answer to two decimal places.

Use the fact that the area of the rectangle = length × width:

$$(x + 4) \times x = 63$$
$$x^2 + 4x = 63$$

Laying out the information in tabular form:

Start with a good guess

Trial value of x	$x^2 + 4x$	Comment
6	$6^2 + (4 \times 6) = 60$	Too low
7	$7^2 + (4 \times 7) = 77$	Too high
So now you know that the value lies between 6 and 7 Try 6.5		
6.5	$6.5^2 + (4 \times 6.5) = 68.25$	Too high
6.3	$6.3^2 + (4 \times 6.3) = 64.89$	Too high
6.2	$6.2^2 + (4 \times 6.2) = 63.24$	Too high
6.1	$6.1^2 + (4 \times 6.1) = 61.61$	Too low
Now you know that the value lies between 6.1 and 6.2 Try 6.15		
6.15	$6.15^2 + (4 \times 6.15) = 62.4225$	Too low
6.18	$6.18^2 + (4 \times 6.18) = 62.9124$	Too low
6.19	$6.19^2 + (4 \times 6.119) = 63.076$	Too high
Now you know that the value lies between 6.18 and 6.19 But you need to give the value to two decimal places Try 6.185		
6.185	$6.185^2 + (4 \times 6.185) = 62.994225$	Too low
Now you know that the value lies between 6.185 and 6.19 But both values are 6.19 to two decimal places		

You can see that the answer is closer to 6 than 7, so you might try 6.3, for example, rather than 6.5

The required answer is $x = 6.19$ to two decimal places.

Apply 1

1 Use the trial and improvement method to solve these equations for a positive value of x.
Show your working and give your answers to one decimal place.
The first one has been started for you.

a $x^2 = 40$

Trial value of x	x^2	Comment
6	36	Too low
7	49	Too high

b $x^2 = 130$ **e** $x^3 = 42$

c $x^2 = 300$ **f** $x^3 = 91$

d $x^2 = 30$ **g** $x^3 = 437$

2 Get Real!

Angus the farmer knows the area of his square fields but not their lengths.
Use the trial and improvement method to find the lengths of the fields.
Give your answers to two decimal places.

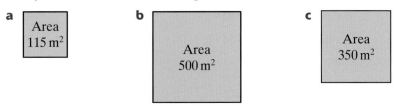

a Area 115 m²

b Area 500 m²

c Area 350 m²

3 Find, using the trial and improvement method,
exact positive solutions of these equations.
Remember to show all your working.

a $x^2 + 2x = 63$ **e** $x^3 + x = 520$

b $x^2 - 2x = 675$ **f** $x^5 = 32\,768$

c $x^2 + 5x = 336$ **g** $x - x^3 = -336$

d $x^2 - 7x = 368$ **h** $2x - x^3 = -711$

4 Get Real!

The area of Sunita's lawn is 50 square metres.
The length of the lawn is one metre longer than the width.
Use the trial and improvement method to work out
the length and width of the lawn.
Give your answers to one decimal place.

5 Use the trial and improvement method to find the lengths of these
rectangles correct to one decimal place.
Remember to show all your working.

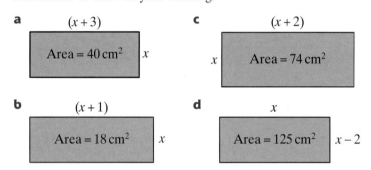

a $(x + 3)$ Area $= 40\,\text{cm}^2$ x

b $(x + 1)$ Area $= 18\,\text{cm}^2$ x

c $(x + 2)$ x Area $= 74\,\text{cm}^2$

d x Area $= 125\,\text{cm}^2$ $x - 2$

6 Use the trial and improvement method to find one solution to each of
these equations.

Give your answers to two decimal places.
Remember to show all your working.

a $x^2 - 2x = 11.4$ if the solution lies between 4 and 5

b $x^3 + x = 616$ if the solution lies between 8 and 9

7 Use the trial and improvement method to find one solution to each of
these equations.

Give your answers to two decimal places.
Remember to show all your working.

a $x^3 = 20$ if x lies between 2 and 3

b $x^2 - 2x + 5 = 23.9$ if x lies between 5 and 6

c $x^3 + 50 = 0$ if x lies between -4 and -3

d $x^3 + 19 = 0$ if x lies between -2 and -3

e $x^3 - 4x + 2 = 0$ if x lies between -2 and -3

f $x^3 + 3x = -176$ if x lies between -5 and -6

Explore

◎ This shape consists of a central square and four equal arms

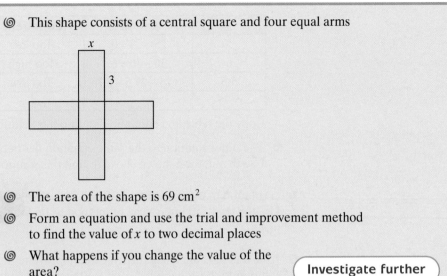

◎ The area of the shape is 69 cm^2

◎ Form an equation and use the trial and improvement method to find the value of x to two decimal places

◎ What happens if you change the value of the area?

Investigate further

Trial and improvement

ASSESS

The following exercise tests your understanding of this chapter, with the questions appearing in order of increasing difficulty.

1 Without using a calculator, decide which of these statements is true and which is false.

a 3.7^2 is between 9 and 16

b 4.6^3 is between 16 and 25

c $\sqrt{55}$ is less than 7

d $\sqrt[3]{55}$ is greater than 3 but less than 4.

2 a Surindra is using a calculator to find the positive square root of 115 by the trial and improvement method. Some of her working is shown below.

x	x²	Comment
10	100	Too low
11	121	Too high
10.5		

Copy the table and continue it to find $\sqrt{115}$ to two decimal places.

b Malachi is using a calculator to find the cube root of 115 by the trial and improvement method. The start of his working is shown below.

x	x³	Comment
4	64	Too low
5		
4.5		

Copy the table and continue it to find $\sqrt[3]{115}$ to two decimal places.

3 a The equation $x^2 - 5x = 4$ has one solution between $x = 5$ and $x = 6$.

x	$x^2 - 5x$	Comment
5	$25 - 25 = 0$	Too low
6	$36 - 30 = 6$	Too high
5.5	$30.25 - 27.5 = 2.75$	Too low
5.75		

Copy the table and continue it to find this solution to two decimal places.

b The other solution to the same equation lies between 0 and −1.
Use the method shown above to find this solution to two decimal places.

4 A solution to the equation $x^3 + 2x = 150$ lies between 5 and 6.
Use the trial and improvement method to find this solution to one decimal place.

5 Josie and Kim are using the trial and improvement method to find a solution to the equation $x^3 - 3x^2 = 10$ to two decimal places.
Josie's answer is 3.72 and Kim's answer is 3.73
Which of them is correct?
Give reasons for your answer.

Try some real past exam questions to test your knowledge:

6 Gary is using the trial and improvement method to find a solution to the equation $x^3 - 5x = 56$

This table shows his first two trials.

x	$x^3 - 5x$	Comment
4	44	Too small
5	100	Too big

Continue the table to find a solution to the equation.
Give your answer to one decimal place.

Spec B, Int Paper 2, Nov 03

7 Dario is using the trial and improvement method to find a solution to the equation $x + \dfrac{1}{x} = 5$

The table shows his first trial.

x	$x + \frac{1}{x}$	Comment
4	4.25	Too low

Continue the table to find a solution to the equation.
Give your answer to one decimal place.

Spec B, Int Paper 2, June 04

Translation and enlargement

D **Examiners would normally expect students who get a D grade to be able to:**

Enlarge a shape by a positive scale factor from a given centre

Translate a shape using a description such as 4 units right and 3 units down

C **Examiners would normally expect students who get a C grade also to be able to:**

Enlarge a shape by a fractional scale factor

Translate a shape by a vector such as $\begin{pmatrix} 4 \\ -3 \end{pmatrix}$

Transform shapes by a combination of translation, rotation, and reflection

Compare the area of an enlarged shape with the original shape

B **Examiners would normally expect students who get a B grade also to be able to:**

Distinguish between formulae for perimeter, area and volume by considering dimensions

A **Examiners would normally expect students who get an A grade also to be able to:**

Compare areas and volumes of enlarged shapes

Enlarge a shape by a negative scale factor

What you should already know ...

- Plot positive and negative coordinates
- Add and subtract negative numbers
- Reflect shapes in a line of symmetry or mirror line
- Rotate shapes around a given centre
- Understand and use units of length

- Give a scale factor of an enlarged shape
- Enlarge a shape by a positive scale factor
- Find the measurements of the dimensions of an enlarged shape

Object – the shape before it undergoes a transformation, for example, translation or enlargement

Image – the shape after it undergoes a transformation, for example, translation or enlargement

Mapping – a transformation or enlargement is often referred to as a mapping with points on the object mapped onto points on the image

Transformation – enlargements and translations are transformations as they change the position but not the size of a shape

Translation – a transformation where every point moves the same distance in the same direction so that the object and the image are congruent

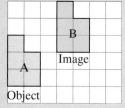

Shape A has been mapped onto Shape B by a translation of 3 units to the right and 2 units up

The vector for this would be $\begin{pmatrix} 3 \\ 2 \end{pmatrix}$

A translation is defined by the distance and the direction (vector)

Vector – used to describe translations

It is written in the form:

$$\begin{pmatrix} \text{units moved in the positive } x\text{-direction} \\ \text{units moved in the positive } y\text{-direction} \end{pmatrix}$$

Congruent – exactly the same size and shape; one of the shapes might be rotated or flipped over

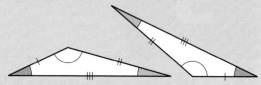

congruent triangles

Similar – figures that are the same shape but different sizes so that one shape may be an enlargement of the other

Scale factor – the ratio of corresponding sides usually expressed numerically so that:

$$\text{Scale factor} = \frac{\text{length of line on the enlargement}}{\text{length of line on the original}}$$

Enlargement – an enlargement changes the size of an object (unless the scale factor is 1) but not its shape; it is defined by giving the centre of enlargement and the scale factor; the object and the image are similar

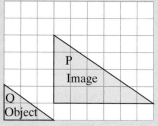

Triangle P is an enlargement of triangle Q
All the lines have doubled in size
The scale factor of the enlargement is 2

Learn 1   Translation

Examples:

Describe the translation that maps triangle P onto triangle Q:

a in words

b as a vector.

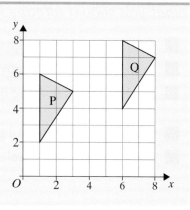

20

The translation that maps P onto Q is:

a 5 units to the right and 2 units up.

b $\begin{pmatrix} 5 \\ 2 \end{pmatrix}$ ← The top number tells you the horizontal movement with positive to the right, negative to the left

The bottom number describes the vertical movement with positive up and negative down

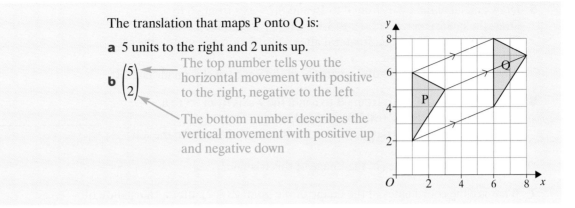

Apply 1

1 Describe the translation that maps triangle X onto triangle Y:

  a in words

  b as a vector.

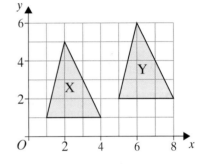

2 a The shaded triangle is translated using the vectors

  $\begin{pmatrix} 3 \\ 2 \end{pmatrix}$ $\begin{pmatrix} -5 \\ 0 \end{pmatrix}$ $\begin{pmatrix} 6 \\ -4 \end{pmatrix}$ $\begin{pmatrix} 5 \\ 4 \end{pmatrix}$

  Write the letters of the images to spell a word.

  b Write the vectors that translate the grey triangle onto the letters of the word FRIEND.

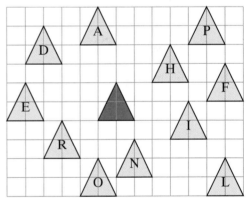

3 Find the coordinates of the images of these points after the translations described.

  a (3, 4) translation 2 units to the right followed by 3 units up.

  b (2, 6) translation 2 units to the left followed by 2 units down.

  c (−2, −3) translation 4 units to the right followed by 6 units up.

  d (4, −1) translation 5 units to the left followed by 3 units down.

4 Find the coordinates of the image of each point after a translation by the given vector.

  a (2, 7) by $\begin{pmatrix} 3 \\ 4 \end{pmatrix}$ **b** (−4, 2) by $\begin{pmatrix} 1 \\ 0 \end{pmatrix}$ **c** (3, −7) by $\begin{pmatrix} -3 \\ -2 \end{pmatrix}$

5 Draw and label the *x*-axis from 0 to 10 and the *y*-axis from −5 to 5.
Draw the quadrilateral A(2, 1), B(3, 4), C(6, 3) and D(7, 1).
Draw the image of the quadrilateral after a translation one unit to the
right and four units down.
What are the coordinates of the vertices of the image of this quadrilateral?

6 Draw and label the *x*-axis from −3 to 6 and the *y*-axis from −5 to 6.
Draw the triangle R(2, 2), S(2, 5) and T(4, 5).
Draw the image of the triangle after a translation three units to the left
and six units down.
What are the coordinates of the image of this triangle?

7 What are the coordinates of the image of the point Z(3, 2) after a translation of:

a $\begin{pmatrix} 3 \\ 1 \end{pmatrix}$ **b** $\begin{pmatrix} 0 \\ -1 \end{pmatrix}$ **c** $\begin{pmatrix} -3 \\ -4 \end{pmatrix}$

8 Triangle PQR is translated to triangle P′Q′R′.
P and Q have coordinates (3, 7) and (5, 8) respectively.
P′ and R′ have coordinates (5, 10) and (4, 1).

Find R and Q′.

9 A translation maps the point L(3, 6) onto the point L¹(−3, 2).
What is the image of the point M(2, 3) under this translation?

10 Triangle A is formed by joining the three points (−2, −2), (−1, −4) and (−1, −1).
Draw this triangle on a pair of axes.

a Show its image, triangle B, after the translation $\begin{pmatrix} 2 \\ -2 \end{pmatrix}$.

Triangle C is found by translating B by the vector $\begin{pmatrix} -2 \\ 2 \end{pmatrix}$.

b What do you notice?

11 F is the point (−2, 1). **X** is the translation $\begin{pmatrix} 1 \\ 3 \end{pmatrix}$. **Y** is the translation $\begin{pmatrix} 2 \\ -5 \end{pmatrix}$.
Z is the translation $\begin{pmatrix} 3 \\ -1 \end{pmatrix}$.

a F is translated by **X** followed by **Y** followed by **Z**.
What are the coordinates of the final image?

b Lauren says that the order of the three translations does not affect the
position of the final image. Is she correct?
Give a reason for your answer.

Explore

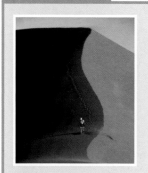

◎ Use a 10 × 10 grid and the triangle X(0, 0), Y(1, 0) and Z(0, 1)

◎ How many different translations are there that use integer values only?
The translated shape must be on the grid

Investigate further

Learn 2 Combining rotation, reflection and translation

A combination of the same transformation or a combination of different transformations can sometimes be described as a single transformation.

Examples:

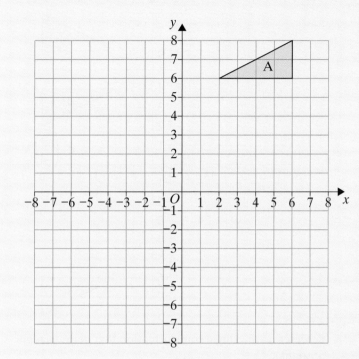

a Triangle A is reflected in the line $y = 5$. Draw this triangle and label it B.

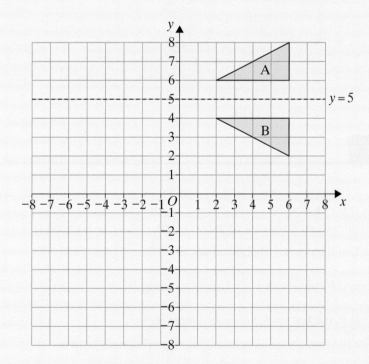

b Triangle B is then translated by $\begin{pmatrix} 0 \\ -10 \end{pmatrix}$ to C. Draw this triangle and label it C.

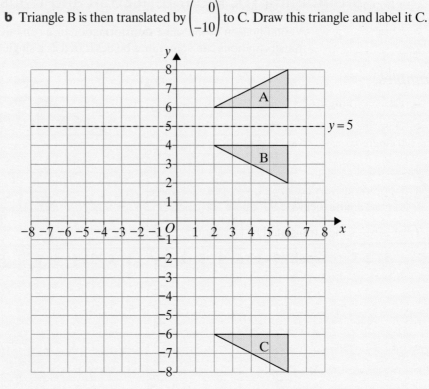

c Describe the single transformation that maps triangle A directly onto triangle C.

Reflection in x-axis $(y = 0)$

Reflection: give the equation the line of reflection
Rotation: give the centre, the angle of rotation and the direction
(clockwise or anticlockwise)
Translation: give the distance and direction, for example, 2 units
to the right and 3 units down (or use vector notation)

Apply 2

1 Draw axes from −8 to 8.
Plot the quadrilateral P with coordinates (−2, 6), (−1, 6), (−1, 1) and (−2, 4).

a Reflect P in the y-axis. Label the image Q.

b Translate Q by the vector $\begin{pmatrix} 2 \\ -4 \end{pmatrix}$. Label the image R.

c Reflect R in the line $x = 4$. Label the image S.

d Describe the single transformation that maps P directly onto S.

2 Draw axes from −8 to 8.
Plot triangle A using the points (6, 0), (4, 3) and (5, 5).

   a Reflect triangle A in the line $x = 4$. Label the image B.

   b Reflect triangle B in the x-axis. Label the image C.

   c Reflect triangle C in the y-axis. Label the image D.

   d Reflect triangle D in the x-axis. Label the image E.

   e Describe the single transformation that maps triangle A directly onto triangle E.

3 Draw axes from −8 to 8.
Triangle X has vertices at (−4, −4), (−4, −8) and (−2, −8).
It is rotated anticlockwise 90° about the point (−2, −2) onto triangle Y.

Triangle Y is then translated by $\begin{pmatrix} 2 \\ 6 \end{pmatrix}$ onto triangle Z.

Describe fully the single transformation that maps triangle X directly onto triangle Z.

4 Plot triangle A with vertices at (1, 1), (2, 1), (2, 3) with axes drawn from −5 to 5.

   a Reflect triangle A in the y-axis and label the image B.

   b Rotate triangle B 90° anticlockwise about the origin and label the image C.

   c Reflect triangle C in the y-axis and label the image D.

   d Describe fully the transformation that will move triangle D back to triangle A.

5 Describe fully three different transformations that could move the square labelled L to the square labelled M.

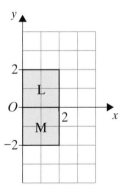

6 Draw the x-axis from −2 to 8 and the y-axis from −8 to 8.
Plot the vertices of the quadrilateral Q at the points (1, 1), (1, 5), (3, 4) and (2, 4).

   a Reflect the quadrilateral Q in the line $y = x$ and label the image R.

   b Reflect R in the x-axis and label the image T.

   c Translate T using the vector $\begin{pmatrix} 2 \\ -4 \end{pmatrix}$ and label the image S.

   d Describe fully the single transformation that maps Q directly onto S.

Learn 3 Enlargement and scale factor

Examples: These two boats are drawn on centimetre grids.

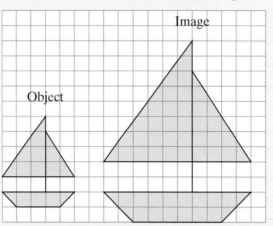

Notice that the two boats are similar
They are the same shape
All corresponding angles are equal
All corresponding lengths are in the same ratio

a What is the scale factor of enlargement?

b What is the area of the small sail on the original drawing (Object)?

c What is the area of the small sail on the enlarged drawing (Image)?

a Scale factor $= \dfrac{\text{length of line on the enlargement}}{\text{length of line on the original}} = \dfrac{6}{3} = 2$ Using length of bottom of boat

b Area of small sail on object $= \frac{1}{2} \times 2 \times 3 = 3 \text{ cm}^2$

c Area of small sail on image $= \frac{1}{2} \times 4 \times 6 = 12 \text{ cm}^2$

Area of enlarged sail $=$ Area of original sail $\times$ (scale factor)^2
$$= 3 \times 2^2$$
$$= 12 \text{ cm}^2$$

Apply 3

1 a Enlarge these shapes by making every line twice as long.

i   **ii**   **iii**

b Work out the areas of the original and the enlarged shapes.

c What do you notice about the areas of the shapes?

2 a Enlarge these shapes by making every line 3 times as long.

i **ii**

b Work out the areas of the original and the enlarged shapes.

c What do you notice about the areas of the shapes?

3 Look at your answers for questions **1** and **2**. What do you notice about the scale factor of the enlargement and the areas of the enlarged shapes when they are compared to the original shapes?

4 Triangle A is enlarged with scale factor 3 into triangle B.
One side of triangle B has length 4.5 cm.

 a What is the length of the corresponding side of triangle A?

One angle of triangle B is 60°.

 b What is the size of the corresponding angle in triangle A?

5 The rectangle ABCD is enlarged by scale factor $\frac{1}{2}$
Find the corresponding lengths of the sides:

 a AB

 b BC

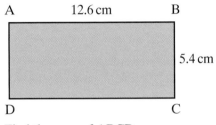

 c Find the area of ABCD.

 d Find the area of the image.

6 Square B is an enlargement of square A, in which a length of 12 cm maps onto a length of 4 cm. Find:

 a the scale factor of enlargement

 b the ratio of the areas.

Learn 4   Centre of enlargement and scale factors

Examples: **a** The triangle XYZ has been enlarged to the triangle QRS.

 i What is the scale factor of the enlargement?

 ii What is the centre of the enlargement?

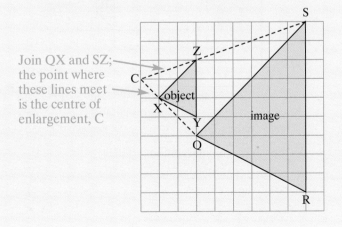

Join QX and SZ; the point where these lines meet is the centre of enlargement, C

The distance from C to Q is 3 times the distance from C to X

The scale factor of enlargement is 3

i The length of YZ is 3 units. The length of RS is 9 units.

$$\text{Scale factor} = \frac{\text{enlarged length}}{\text{corresponding original length}} = \frac{9}{3} = 3$$

Scale factor = 3

ii Point C is the centre of the enlargement.

> To define an enlargement, give the centre of the enlargement and the scale factor

b The quadrilateral ABCD is enlarged to A'B'C'D'. Find:

i the scale factor

ii the centre of enlargement.

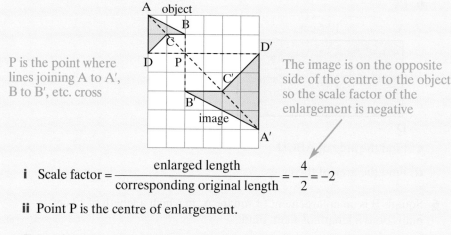

> P is the point where lines joining A to A', B to B', etc. cross

> The image is on the opposite side of the centre to the object so the scale factor of the enlargement is negative

i $\text{Scale factor} = \dfrac{\text{enlarged length}}{\text{corresponding original length}} = -\dfrac{4}{2} = -2$

ii Point P is the centre of enlargement.

c Trapezium B is an enlargement of trapezium A. Find:

i the scale factor

ii the centre of enlargement.

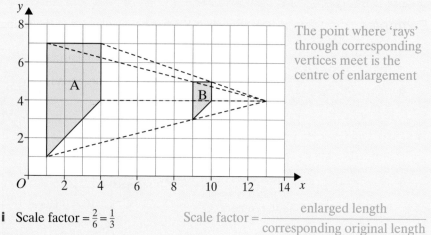

> The point where 'rays' through corresponding vertices meet is the centre of enlargement

i Scale factor $= \frac{2}{6} = \frac{1}{3}$

$\text{Scale factor} = \dfrac{\text{enlarged length}}{\text{corresponding original length}}$

ii The centre of enlargement = (13, 4)

An enlargement by a scale factor between zero and one will make the shape smaller

Apply 4

1 Copy each diagram and draw an enlargement with scale factor 2.
   Use point C as the centre of the enlargement.

   a 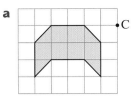   **b**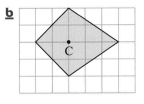

2 For each of these, copy the diagram and draw an enlargement with scale
   factor 3. Use point C as the centre of enlargement.

   a 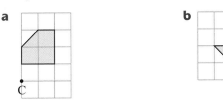   **b**

3 Enlarge the L-shape from centre $(0, 0)$ with scale factor $\frac{1}{2}$

   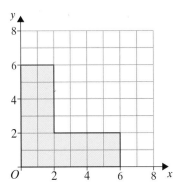

4 Draw and label the x-axis from 0 to 18 and the y-axis from 0 to 20.
   Draw each of these shapes and enlarge it as indicated.

   a Rectangle $(4, 6)$, $(12, 14)$, $(8, 18)$, $(0, 10)$; with centre of enlargement $(8, 14)$
   and scale factor $\frac{1}{4}$

   b Trapezium $(1, 2)$, $(3, 2)$, $(3, 3)$, $(1, 4)$; with centre of enlargement $(5, 3)$
   and scale factor -3

5 Draw and label the x-axis from -6 to 3 and the y-axis from -4 to 2.
   Enlarge the shape made by joining the points $(1, 2)$, $(3, 2)$, $(3, 1)$ and
   $(1, 1)$ from the centre $(0, 0)$ with a scale factor of -2.

6 In each of these diagrams, shape X has been enlarged to make shape Y.
In each case find the scale factor and the centre of the enlargement.

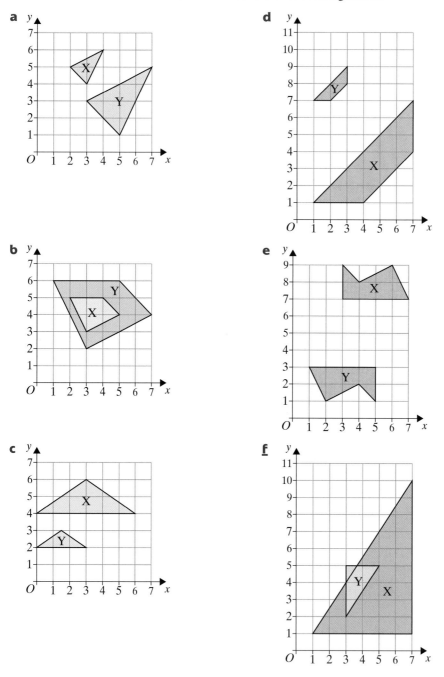

7 Draw and label the *x*-axis from 0 to 12 and the *y*-axis from 0 to 10.
Plot and label pentagon P with vertices at $(7, 1)$, $(11, 1)$, $(10, 5)$,
$(11, 7)$ and $(7, 7)$.

   a Draw the image of pentagon P after enlargement scale factor $\frac{1}{2}$, centre
$(1, 9)$. Label the image Q.

   b What are the coordinates of the vertices of Q?

8 Draw and label the *x*-axis from −8 to 3 and the *y*-axis from −10 to 4.
Plot and label the triangle A with vertices at (1, 1), (1, 3) and (2, 1).

   a Draw the image of triangle A after enlargement scale factor −3,
centre (0, 0). Label the image B.

   b What are the coordinates of the vertices of B?

9 Draw and label the *x*-axis from −8 to 4 and the *y*-axis from −4 to 5.
Plot and label the rectangle R with vertices at (−6, 4), (−1, 4), (−1, 1)
and (−6, 1).

   a Draw the image of rectangle R after enlargement scale factor $-\frac{1}{2}$
centre (0, 0). Label the image S.

   b What are the coordinates of the vertices of S?

Explore

  ◎ Draw a shape of your choice

  ◎ Enlarge the shape, using centres of enlargement both inside and outside
the shape

  ◎ What happens when you use different scale factors (whole numbers,
negative numbers ...)?

Investigate further

Learn 5   Enlargement and ratio

Examples: **a** Rectangle B is enlarged to give rectangle A.

      i What is the ratio of the lengths and widths of their sides?

      ii What is the ratio of the perimeter of B to the perimeter of A?

      iii What is the ratio of the area of B to the area of A?

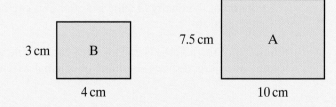

i In rectangle B the length is 4 cm. In rectangle A the length is 10 cm.
The ratio of the lengths is 4 : 10. This can be simplified to 2 : 5.

In rectangle B the width is 3 cm.
In rectangle A the width is 7.5 cm.
The ratio of their widths is 3 : 7.5.
This can be simplified to 6 : 15 = 2 : 5.

The ratios of all of the sides should be the same because A is an enlargement of B, scale factor of 2.5. The two shapes are similar

ii The perimeter of rectangle B is 2(3 + 4) = 14 cm.
The perimeter of rectangle A is 2(7.5 + 10) = 35 cm.
The ratio of the perimeter of B to A = 14 : 35.
This can be simplified to become 2 : 5. ◄——— *Perimeter is a length also, so the ratio should be the same as in part **a***

iii The area of rectangle B is 3 × 4 = 12 cm^2.
The area of rectangle A is 7.5 × 10 = 75 cm^2.
The ratio of the area of B to A = 12 : 75.
This can be simplified to become 4 : 25

The ratio of the lengths is 2 : 5. The ratio of the areas is 4 : 25. ($2^2 : 5^2$)

If the ratio of the lengths is $x : y$ then the ratio of the areas is $x^2 : y^2$
Area of enlarged shape = area of original × (scale factor)^2

b A cube has edges of 2 cm. It is enlarged so that all of its edges are 4 cm long.

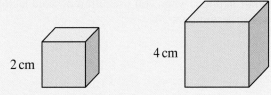

2 cm   4 cm

Find:

   i the ratio of the lengths

  ii the ratio of the volumes.

   i The length of the original edge is 2 cm.
The length of the enlarged edge is 4 cm.
The ratio of the original edge to the enlarged edge is 2 : 4.
This can be simplified to 1 : 2.

  ii The volume of the original cube is 2 × 2 × 2 = 8 cm^3.
The volume of the enlarged cube is 4 × 4 × 4 = 64 cm^3.
The ratio of the volume of the original cube to the
volume of the enlarged cube is 8 : 64. This can be
simplified to 1 : 8.

The ratio of the lengths is 1 : 2. The ratio of the volumes is 1 : 8. ($1^3 : 2^3$)

If the ratio of the lengths is $x : y$ then the ratio of the volumes is $x^3 : y^3$
Volume of enlarged shape = volume of original × (scale factor)^3

Apply 5

1 Rectangle D is an enlargement of rectangle C.

a What is the ratio of the lengths of these rectangles (in its simplest form)?

b What is the ratio of the widths of these rectangles (in its simplest form)?

c What is the ratio of the perimeter of these rectangles (in its simplest form)?

d What is the ratio of the areas of these rectangles (in its simplest form)?

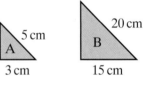

2 Triangle F is an enlargement of the isosceles triangle E.

a What is the ratio of the base of these triangles (in its simplest form)?

b What is the ratio of the sides of these triangles (in its simplest form)?

c What is the ratio of the perimeter of these triangles (in its simplest form)?

d What is the ratio of the areas of these triangles (in its simplest form)?

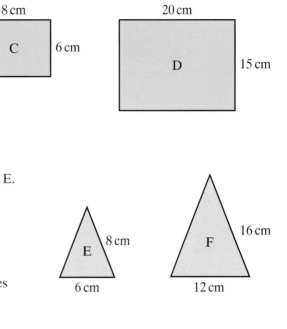

3 James says that triangle B is an enlargement of triangle A. Is he correct? Give a reason for your answer.

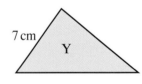

4 Triangle Y is an enlargement of triangle X.

a Work out the ratio of the corresponding lengths.

b Calculate the lengths of the unknown sides.

c What is the ratio of the perimeters of these triangles (in its simplest form)?

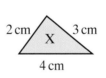

5 Trapezium S is an enlargement of trapezium T.

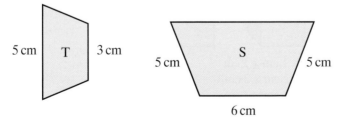

a Work out the ratio of corresponding lengths.

b Calculate the lengths of the unknown sides.

c What is the ratio of the perimeters of these trapeziums?

6 These pairs of solids are similar. State the ratio of the corresponding lengths. Calculate the lengths of w, x and y.

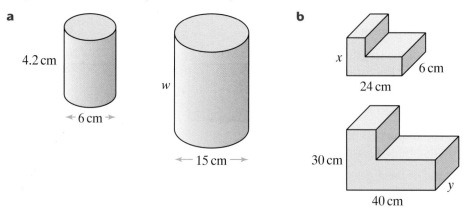

a

4.2 cm

6 cm

w

15 cm

b

x

24 cm

6 cm

30 cm

40 cm

y

7 The area of a shape is 4 cm². It is enlarged by a scale factor 3. What is the area of the enlarged shape?

8 The area of a shape is 15 cm². It is enlarged by a scale factor 1.5 What is the area of the enlarged shape?

9 A cube of 2 cm is enlarged so that its edges are 10 cm long. What is the ratio of the volumes?

10 Two similar bottles have heights of 25 cm and 50 cm. Find:

   a the ratio of the volumes of the bottles

   b the volume (in litres) of the larger bottle if the smaller one holds 200 mℓ.

11 Two similar cylinders are 15 cm and 60 cm high. The small cylinder holds 500 mℓ. What is the capacity of the larger cylinder?

12 Cooking oil is sold to a catering firm in two similar sizes of containers. The volume of these containers is 2 litres and 54 litres. What is the height of the larger container if the smaller one is 25 cm high?

Explore

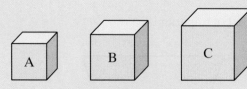

A   B   C

◎   Cube A is 1 cm × 1 cm × 1 cm
Cube B is 2 cm × 2 cm × 2 cm
Cube C is 3 cm × 3 cm × 3 cm

◎   Write down the ratio of the lengths of the cubes

◎   Write down the ratio of the volumes of the cubes

◎   Is there a relationship between your answers?

Investigate further

Learn 6 Dimensions of formulae

Examples:

The letters a, b and c each represent a length.
State whether each of these formulae could represent length, area, volume or makes no sense.

a $a + b$ **d** abc

b $4\pi ab$ **e** $2ab^2$

c $2\pi a^2$ **f** $ab + c$

a $a + b$ a represents a length, b represents a length

length + length   ⟵ Length is one-dimensional

= length

b $4\pi ab$ a and b are both lengths

length × length

= length^2

= area

A symbol or a number that does not represent a number of units of length, area or volume has no effect on the dimension of an expression

Length × length is two-dimensional

c $2\pi a^2$ a is length, a^2 represents $a \times a$

length × length   Ignore 2 and π as they do not have dimensions

= length^2

= area

d abc a represents a length, b represents a length, c represents a length

length × length × length

= length^3 Length × length × length is three-dimensional

= volume

e $2ab^2$ a represents a length, b represents a length, b^2 represents $b \times b$

length × length × length   Ignore 2 as it does not have dimensions

= length^3

= volume

f $ab + c$ ab represents area, c represents length

area + length – this makes no sense as area cannot be added to length

Length is measured in units (one-dimensional)

Area is measured in units^2 (two-dimensional)

Volume is measured in units^3 (three-dimensional)

Apply 6

1 State whether each of these quantities is a length, an area or a volume.

a 8 cm^2 **e** 9 mm **i** 120 mm^3

b 6 cm **f** $2\pi \text{ cm}^2$ **j** 21 km

c 12 cm^2 **g** 16 km^3 **k** 37 m^3

d 5 cm^3 **h** 4 m **l** 45 m^2

2 The letters l, h, r, b, w, d each represent a length.
State whether each of these formulae could represent length, area or volume.

a πr^2 **e** $4(l^2 + wd)$ **i** $5(l + h)^2 d$

b $l + h$ **f** $2(w + h)^2$ **j** $\dfrac{rbw}{rb}$

c πd **g** $\pi r^2 h$

d lhr **h** $b^2 + d^2$

3 Lauren was asked to find the area of a circle with a diameter of 12 cm.
She wrote down:

    Area $= \pi$d
        $= \pi \times 12$
        $= 37.7 \text{ cm}^2$

Vicky could not remember the formula for the area of a circle but she
knew that Lauren was wrong. How did she know?

4 For these formulae, decide if X could represent a length, an area,
a volume or makes no sense.
a, b, h, l, r are all lengths.

a $X = 2ab + hl$ **e** $X = 5lr + a$ **i** $X = ahl + rb$

b $X = 2\pi r^3$ **f** $X = \dfrac{a^2}{b}$ **j** $\dfrac{a + b}{a}$

c $X = \frac{1}{3}\pi r^2 h$ **g** $X = 2abhl$

d $X = 4(l + b)$ **h** $X = r^2 b$

5 In these formulae A represents area, V represents volume and b, d, h, l
and r represent length. State whether each of these formulae could
represent length, area, volume or makes no sense.

a $\dfrac{b^2 + d^2}{h}$ **d** $V + rbd$ **g** $\frac{1}{2}A$

b Ah **e** $\dfrac{V}{A}$ **h** $V + r^2$

c $2\sqrt{A + l^2}$ **f** $\dfrac{l^3}{h}$ **i** $\dfrac{hlr^2}{3bd}$

6 The letters x, y and z all represent lengths.
From this list of expressions write those that could represent:

a length   **b** area   **c** volume   **d** none of these.

i $x + y + z$ **v** $\frac{1}{2}xy$

ii $x^2 + 2xyz$ **vi** $\sqrt{x^2yz}$

iii $\dfrac{yz}{x} + \sqrt{xz}$ **vii** $\dfrac{xyz}{3}$

iv $\pi z^3 + 2xy^2$

7 Some of these formulae have an error.
Explain how you can tell which ones are wrong.
Suggest a correct formula, if possible.
A represents area, V represents volume and b, d, h, l and r represent lengths.

a $A = bd$ **c** $A = \pi l^2 r$ **e** $V = d^2 + h^3$

b $V = \pi r^2 h^2$ **d** $h = 2Ab$ **f** $A = b(l + r)$

8 a, b and c are lengths. Combine a, b and c to form five formulae that could represent:

a volume   **b** area.

Translation and enlargement

ASSESS

The following exercise tests your understanding of this chapter, with the questions appearing in order of increasing difficulty.

1 Use squared paper to copy and draw enlargements of these shapes with the given scale factors from the given centres of enlargement O.

a

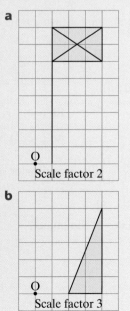

Scale factor 2

b

Scale factor 3

c

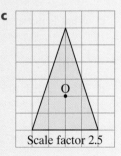

Scale factor 2.5

2 Use squared paper to copy and draw enlargements of these shapes with the given scale factors from the given centres of enlargement O.

a

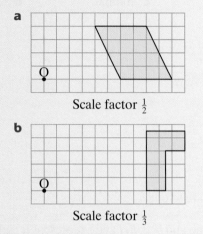

Scale factor $\frac{1}{2}$

b

Scale factor $\frac{1}{3}$

3 a A walker needs to get to the other side of a lake.
To do this she must first walk 4 km due west and then 3 km due north. Draw her journey on squared paper and then measure and write down the actual distance she is from her original starting place.

b Write down the coordinates of the images of each of the following points after the given translations:

i Object (2, 7), Translation $\begin{pmatrix} 2 \\ 1 \end{pmatrix}$

ii Object (−3, 9), Translation $\begin{pmatrix} 4 \\ -3 \end{pmatrix}$

iii Object (2, −7), Translation $\begin{pmatrix} -5 \\ -1 \end{pmatrix}$

iv Object (−4, −6), Translation $\begin{pmatrix} -6 \\ 2 \end{pmatrix}$

c What is noticeable about the following translations?

i $\begin{pmatrix} 3 \\ 0 \end{pmatrix}$ **ii** $\begin{pmatrix} 0 \\ -2 \end{pmatrix}$

d What translation transforms:

i (3, 0) into (7, −2) **iii** (−4, −7) into (4, 7)

ii (5, −2) into (2, 4) **iv** (−2, 3) into (2, 3)?

e After a translation of $\begin{pmatrix} -3 \\ 4 \end{pmatrix}$, the coordinates of a square are (4, 3), (−1, 3), (4, −2) and (−1, −2).

What are the coordinates of the object?

4 Draw and label axes for values of x and y between -16 and 16.

  a On the diagram draw a letter F by plotting and joining the points $(2, 1)$, $(2, 3)$, $(2, 5)$, $(3, 3)$ and $(4, 5)$.

  b Enlarge the letter F by scale factor 3, centre $(0, 0)$.

  c Rotate your enlargement by $90°$ anticlockwise about $(0, 0)$.

  d Translate this rotation by the vector $\begin{pmatrix} 10 \\ -13 \end{pmatrix}$.

5 a Enlarge the given shape from centre (O) with scale factor $-\frac{1}{2}$

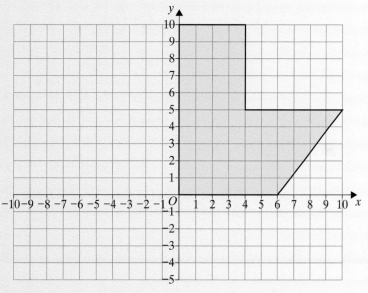

  b Find the centre and scale factor of the enlargements shown.

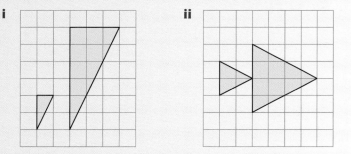

    i   ii

6 a, b and c are different lengths. State whether each of these formulae could represent a length, an area, a volume, or none of these.

  a $2ab$ e $4c + b^2$

  b $a - b$ f $5b^2c$

  c $2(a + b)$ g $\dfrac{4ab}{c}$

  d $\dfrac{4a}{b}$ h $b^2 + c^3$

7 a A frame encloses an area of 3 m^2.
A similar frame has dimensions twice that of the original.
What is the area enclosed by the larger frame?

   b A traffic policeman finds that it takes a motorist 3 seconds to completely fill a breathalyser bag. How long would it take the same motorist, blowing at the same rate, to fill a similar bag twice as big?

Try a real exam question to test your knowledge:

8

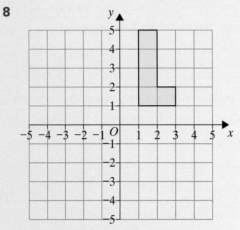

Enlarge the shaded shape by scale factor $-\frac{1}{2}$ with centre of enlargement $(-1, 0)$.

Spec A, Higher Paper 2, June 04

4 Measures

OBJECTIVES

C **Examiners would normally expect students who get a C grade to be able to:**

Solve more difficult speed problems

Understand and use compound measures such as speed and density

Recognise accuracy in measurements given to the nearest whole unit

B **Examiners would normally expect students who get a B grade also to be able to:**

Find the upper and lower bounds of simple calculations involving quantities given to a particular degree of accuracy

A–A* **Examiners would normally expect students who get an A–A* grade also to be able to:**

Find the upper and lower bounds of more difficult calculations with quantities given to various degrees of accuracy

What you should already know ...

- Decide which metric unit to use for everyday measurements
- Convert between imperial and metric units
- Know rough metric equivalents of pounds, feet, miles, pints and gallons
- Make sensible estimates of a range of measures in everyday settings
- Rounding numbers

VOCABULARY

Unit – a standard used in measuring, for example, a metre is a unit of length

Metric units – these are related by multiples of 10 and include:
- metres (m), millimetres (mm), centimetres (cm) and kilometres (km) for lengths
- grams (g), milligrams (mg), kilograms (kg) and tonnes (t) for mass
- litres (ℓ), millilitres (mℓ) and centilitres (cℓ) for capacity

Imperial units – these are units of measurement historically used in the United Kingdom and other English-speaking countries; they are now largely replaced by metric units. Imperial units include:
- inches (in), feet (ft), yards (yd) and miles for lengths
- ounces (oz), pounds (lb), stones and tons for mass
- pints (pt) and gallons (gal) for capacity

Conversion factor – the number by which you multiply or divide to change measurements from one unit to another. The approximate conversion factors that you should know are:

Length	Mass	Capacity
1 foot ≈ 30 cm	1 kg ≈ 2.2 pounds	1 gallon ≈ 4.5 litres
5 miles ≈ 8 km		1 litre ≈ 1.75 pints

The table below gives conversion factors for metric units of length, mass and capacity

Metric system ~~Learn these facts~~

Length	Mass	Capacity
1 cm = 10 mm	1 g = 1000 mg	1 ℓ = 100 cℓ or 1000 mℓ
1 m = 100 cm or 1000 mm	1 kg = 1000 g	
1 km = 1000 m	1 t = 1000 kg	

Compound measure – a measure formed from two or more other measures, for example,

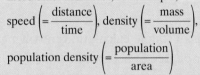

$$\text{speed} \left(= \frac{\text{distance}}{\text{time}}\right), \text{density} \left(= \frac{\text{mass}}{\text{volume}}\right),$$

$$\text{population density} \left(= \frac{\text{population}}{\text{area}}\right)$$

Speed – the rate of change of distance with respect to time. To calculate the average speed, divide the total distance moved by the total time taken. It is usually given in metres per second (m/s) or kilometres per hour (km/h) or miles per hour (mph)

In the triangle, cover the item you want, then the rest tells you what to do

$$\text{Speed} = \frac{\text{distance}}{\text{time}} \qquad \frac{\text{metres}}{\text{seconds}} \text{ gives m/s}$$

Distance = speed × time

$$\text{Time} = \frac{\text{distance}}{\text{speed}}$$

Density – to calculate density, divide the mass of the object by the volume of the object. It is usually given in grams per cubic centimetre (g/cm^3) or kilograms per cubic metre (kg/m^3)

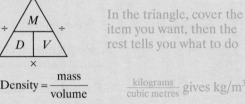

In the triangle, cover the item you want, then the rest tells you what to do

$$\text{Density} = \frac{\text{mass}}{\text{volume}} \qquad \frac{\text{kilograms}}{\text{cubic metres}} \text{ gives kg/m}^3$$

Mass = density × volume

$$\text{Volume} = \frac{\text{mass}}{\text{density}}$$

Lower bound – this is the minimum possible value of a measurement, for example, if a length is measured as 37 cm correct to the nearest centimetre, the lower bound of the length is 36.5 cm

Upper bound – this is the maximum possible value of a measurement, for example, if a length is measured as 37 cm correct to the nearest centimetre, the upper bound of the length is 37.5 cm

Learn 1   Compound measures

Examples:

a A swimmer crosses a river that is 150 metres wide in $7\frac{1}{2}$ minutes.

    i Find her average speed in metres per second and kilometres per hour.

    ii Estimate her average speed in miles per hour.

b The density of gold is 19.3 g/cm^3.

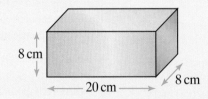

A gold bar is 20 cm long, 8 cm wide and 8 cm high.

    i How much does it weigh to the nearest kilogram?

    ii Estimate its mass in pounds.

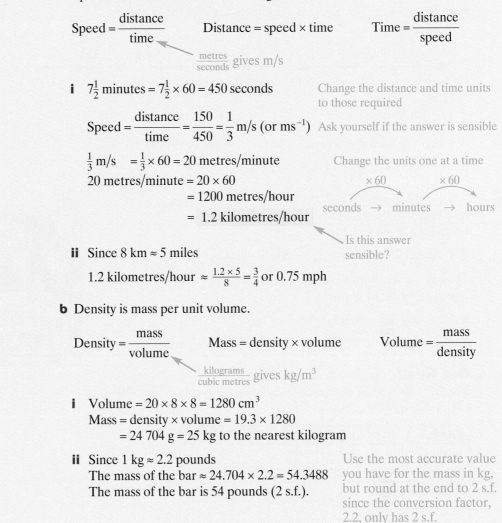

a Speed is the rate at which something moves.

$$\text{Speed} = \frac{\text{distance}}{\text{time}} \qquad \text{Distance} = \text{speed} \times \text{time} \qquad \text{Time} = \frac{\text{distance}}{\text{speed}}$$

$\frac{\text{metres}}{\text{seconds}}$ gives m/s

i $7\frac{1}{2}$ minutes $= 7\frac{1}{2} \times 60 = 450$ seconds

Change the distance and time units to those required

$$\text{Speed} = \frac{\text{distance}}{\text{time}} = \frac{150}{450} = \frac{1}{3} \text{ m/s (or ms}^{-1}\text{)}$$

Ask yourself if the answer is sensible

$\frac{1}{3}$ m/s $= \frac{1}{3} \times 60 = 20$ metres/minute

Change the units one at a time

20 metres/minute $= 20 \times 60$
$= 1200$ metres/hour
$= 1.2$ kilometres/hour

$\times 60 \qquad \times 60$

seconds $\rightarrow$ minutes $\rightarrow$ hours

Is this answer sensible?

ii Since 8 km $\approx$ 5 miles

1.2 kilometres/hour $\approx \frac{1.2 \times 5}{8} = \frac{3}{4}$ or 0.75 mph

b Density is mass per unit volume.

$$\text{Density} = \frac{\text{mass}}{\text{volume}} \qquad \text{Mass} = \text{density} \times \text{volume} \qquad \text{Volume} = \frac{\text{mass}}{\text{density}}$$

$\frac{\text{kilograms}}{\text{cubic metres}}$ gives kg/m^3

i Volume $= 20 \times 8 \times 8 = 1280$ cm^3
Mass $=$ density $\times$ volume $= 19.3 \times 1280$
$= 24\,704$ g $= 25$ kg to the nearest kilogram

ii Since 1 kg $\approx$ 2.2 pounds
The mass of the bar $\approx 24.704 \times 2.2 = 54.3488$
The mass of the bar is 54 pounds (2 s.f.).

Use the most accurate value you have for the mass in kg, but round at the end to 2 s.f. since the conversion factor, 2.2, only has 2 s.f.

Apply 1

1 a The steam engine, Mallard, took 2 hours to travel 253 miles. What was its average speed?

b How far will a ball roll in 12 seconds at an average speed of 2.5 m/s?

c How long did Kath take to run the 400 metre race if her average speed was 8 m/s?

d Ben cycled for 20 minutes at a steady speed of 18 mph. How far did he travel?

e Eurostar travelled 330 kilometres from London to Brussels in 2 hours 20 minutes. What was its average speed?

f Approximately how long will it take a walker to travel 3 miles at a steady speed of 2 m/s?

2 Get Real!

The diagram gives the times that a train stops at stations on a journey
from Newcastle to Derby and the distances between these stations.

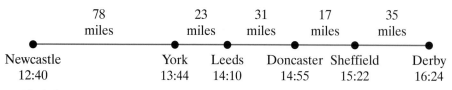

	78 miles		23 miles	31 miles	17 miles	35 miles	
Newcastle 12:40		York 13:44	Leeds 14:10	Doncaster 14:55	Sheffield 15:22	Derby 16:24	

a Find the average speed:

i between each station and the next   **ii** for the whole journey.

b Compare and comment on your answers.

3 A delivery van sets off at 7:45 a.m. to deliver a package to a
customer who lives 20 miles from the store. It arrives at the
customer's house at 8:10 a.m.

a Calculate the van's average speed in miles per hour.

The van's average speed on the return journey in rush hour traffic
is only 25 miles per hour.

b How long does the return journey take?

c Calculate the van's average speed for the whole journey.

4 Get Real!

The table gives some of the results from men's events in the 2004
Olympic Games.

Winner	Distance	Time
Justin Gatlin	100 m	9.85s
Shawn Crawford	200 m	19.79s
Jeremy Wariner	400 m	44.00s
Yuriy Borzakovskiy	800 m	1min 44.45s
Hicham El Guerrouj	1500 m	3min 34.18s
Hicham El Guerrouj	5000 m	13min 14.39s
Kenenisa Bekele	10000 m	27min 05.10s

Find the average speed of each runner in:

a metres per second

b kilometres per hour.

5 Five students are told that an animal travels half a kilometre in 25 minutes.
The answers they give for the average speed are:

Natasha 12.5 m/s   Sue 20 m/s   Sam 50 m/s   Isaac 33 cm/s   Mike 3 m/s

a Who gave the correct answer?

b Explain what you think each of the other students did wrong.

6 The average speed for a journey was 2 m/s.
Give a possible distance and time:

a in metres and seconds   **b** in kilometres and minutes.

7 A driver plans to start a journey of 210 miles at 10 a.m.

    a He expects to travel at an average speed of 60 mph.
What time does he expect to arrive?

    b If his car's rate of petrol consumption is 7.5 miles per litre,
calculate the amount of petrol he will use:

        i in litres   **ii** in gallons (to the nearest gallon).

8 Get Real!

    a i Find the distance you travel to school.

      ii Time how long this journey takes on a number of occasions.

    b Work out your average speed on each journey and your overall average speed.

9 Get Real!

The table gives the mean radius of the orbit of each planet in the solar
system and the time taken by each planet to go around the sun.

Planet	Mean radius (million km)	Time taken
Mercury	58	88 days
Venus	108	226 days
Earth	150	1.00 years
Mars	228	1.88 years
Jupiter	778	11.86 years
Saturn	1427	29.46 years
Uranus	2871	84.01 years
Neptune	4497	164.79 years
Pluto	5914	248.54 years

    a Assuming that each orbit is approximately circular, find the speed of each
planet to the nearest thousand kilometres per hour.

    b What can you say about the way the speed of a planet depends on its
distance from the sun?

10 Copy and complete the table.

Mass	Volume	Density
248 g	200 cm^3	
4.5 kg	0.6 m^3	
	16 cm^3	4.5 g/cm^3
	1.2 m^3	640 kg/m^3
456 g		0.8 g/cm^3
7.2 kg		1200 kg/m^3

11 The density of petrol is 737 kg/m^3.
What is the mass of the petrol in a 100 litre tank when it is full?
Give your answer:

    a in kilograms   **b** in pounds.

12 A cylindrical log of wood weighs 712 kg.
It is 2.5 m long and has a diameter of 80 cm.
Find its density:

    a in kg/m^3 **b** in g/cm^3.

13 The density of copper is $8.9 \ g/cm^3$ and the density of tin is $7.3 \ g/cm^3$.
400 g of copper and 600 g of tin are melted together to form an alloy.
Find the density of the alloy.

14 Find the value of the missing quantities in each part:

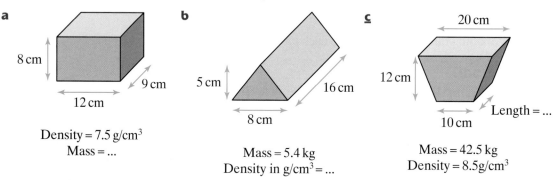

  a

8 cm   9 cm   12 cm

Density = $7.5 \ g/cm^3$
Mass = ...

  b

5 cm   16 cm   8 cm

Mass = 5.4 kg
Density in g/cm^3 = ...

  c

20 cm   12 cm   Length = ...   10 cm

Mass = 42.5 kg
Density = $8.5 g/cm^3$

15 The Earth has mass 5.97×10^{21} tonnes and average density $5520 \ kg/m^3$.
Find its volume in km^3.

16 A plaque consists of a bronze plate in the shape of
a regular hexagon standing on a wooden base in the
shape of a cuboid. Its dimensions are given in the
diagram.
The density of bronze is $8.7 \ g/cm^3$ and the density
of wood is $0.57 \ g/cm^3$.

Calculate how much the plaque weighs:

  a in kilograms   **b** in pounds.

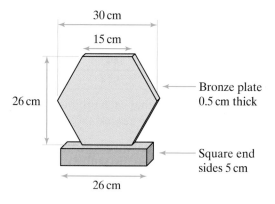

30 cm

15 cm

26 cm

Bronze plate
0.5 cm thick

Square end
sides 5 cm

26 cm

17 Population density is the number of people per square kilometre who
live in an area.

  a A town covers an area of 16 square kilometres and has a population
of 35 360.
What is its population density?

  b It is planned to extend the town by building a new housing estate on
adjacent land of area 0.25 square kilometres. The plan assumes the
population density of the estate will be 4500 people per square kilometre.
According to the plan, how many people will live on the estate?

  c Find the population density of the town including the new estate:

    i assuming that no new people move into the area and that the
houses on the estate are all occupied by people who already live
in the town

    ii assuming that the houses on the estate are all occupied by
newcomers into the area.

Explore

◎ Find the diameter, thickness and weight of a 10p coin
You will get a more accurate result if you weigh a number of coins

◎ Calculate the density of the coin

Investigate further

Learn 2 Calculating with minimum and maximum values

Example:

A sprinter runs 100 m in 13 seconds. Both measurements are rounded to the nearest whole unit. What is the maximum possible speed of the runner in this 100 m race?

Remember that measurements given to the nearest whole unit may be up to half a unit larger or smaller. When you use them in calculations the possible errors in the measurements may affect the results.

The following inequality statements show the lower and upper bounds of the distance and the time:

To get the *biggest* possible answer, divide by the *smallest* possible time

$$99.5 \text{ m} \leqslant \text{distance} < 100.5 \text{ m}$$
$$12.5 \text{ s} \leqslant \text{time} < 13.5 \text{ s}$$

$$\text{Speed} = \frac{\text{distance travelled}}{\text{time taken}}$$

$$\text{Maximum possible speed} = \frac{\text{maximum possible distance travelled}}{\text{minimum possible time taken}}$$

$$\text{So maximum possible speed} = \frac{100.5 \text{ m}}{12.5 \text{ s}} = 8.04 \text{ m/s}$$

Minimum possible distance

$$\text{Note that the minimum possible speed} = \frac{99.5 \text{ m}}{13.5 \text{ s}} = 7.37 \text{ m/s}$$

Maximum possible time

To find the lower and upper bounds of a combination of two measurements, use the following:

$$\text{minimum } (x + y) = \text{minimum } x + \text{minimum } y$$
$$\text{minimum } (x - y) = \text{minimum } x - \textbf{maximum } y$$
$$\text{minimum } (x \times y) = \text{minimum } x \times \text{minimum } y$$
$$\text{minimum } (x \div y) = \text{minimum } x \div \textbf{maximum } y$$

$$\text{maximum } (x + y) = \text{maximum } x + \text{maximum } y$$
$$\text{maximum } (x - y) = \text{maximum } x - \textbf{minimum } y$$
$$\text{maximum } (x \times y) = \text{maximum } x \times \text{maximum } y$$
$$\text{maximum } (x \div y) = \text{maximum } x \div \textbf{minimum } y$$

Apply 2

1 Calculate the maximum possible speed for a journey of 78 miles in 2.1 hours, where both measurements are given correct to two significant figures.

2 Anne drives 35.3 miles in 46.5 minutes. Calculate Anne's maximum possible speed, if the numbers are given correct to three significant figures.

3 Joe drove 35 miles to a football match, then 10 miles to drop his friend off, then 30 miles back home. All these distances are correct to the nearest 5 miles. What is the minimum possible distance that Joe drove?

4 Get Real!

The River Nile is 6825 miles long and the River Amazon is 6435 miles long. If each of these measurements is correct to the nearest 5 miles, what is the maximum and minimum difference between their lengths?

5 A ball of string is 5 metres long to the nearest 10 cm. If it is cut into two identical pieces, write the length of each piece in the form $a \leqslant$ length of piece $< b$.

6 A pack of butter weighs 200 grams to the nearest gram.
Find upper and lower bounds for the total mass of 6 packs of butter.

7 A teacher gives his students the following question:
'A cheese weighs 2 kg to the nearest 50 g. It is cut into pieces weighing 250 g to the nearest 10 g. Find the maximum possible number of pieces.'
The working of three students is given below.

John's working:	Hazel's working:	Farid's working:
$2050 \div 240 = 8.54 \ldots$	$2025 \div 255 = 7.94 \ldots$	$2025 \div 245 = 8.26 \ldots$
Answer: 9 pieces	Answer: 7 pieces	Answer: 8 pieces

a Who is correct?

b Explain the mistakes each of the others has made.

8 The radius of a circle is 4.5 cm correct to 1 decimal place.

a Find upper and lower bounds for its area.

b What difference does it make if you use 3.14 instead of the π key?

9 p, q and r are three measurements. In each case, say whether you will need to use the maximum possible or the minimum possible value of each measurement in order to find the:

a minimum value of $p + q + r$

b maximum value of $p - q$

c maximum value of $\dfrac{p - q}{r}$

10 A cuboid has length 26 cm, width 17 cm and height 13 cm, all correct to the nearest centimetre.

    a Calculate the upper and lower bounds of the volume of the cuboid.

    b To what accuracy is it possible to give the volume of the cuboid?

    c What is the minimum possible length of the diagonal of the cuboid?

11 The density of wood is 0.71 g/cm^3, correct to 2 significant figures.
A block of wood weighs 459 g, correct to the nearest gram.
What is the upper bound of the volume of the block of wood?

12 Sameera runs 100 m (to the nearest 10 m) in 12 seconds (correct to the nearest second).
Last time she ran the race, her maximum possible speed was 8.2 metres per second.
Can she be sure that she has run faster this time?
Give a reason for your answer.

Explore

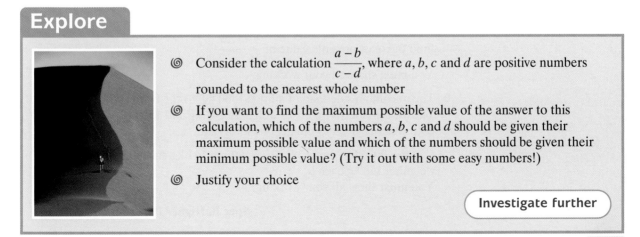

The diagram shows the length and width of a room

    4 m

    6 m

◎ Assuming that the measurements are correct to the nearest metre, calculate
   – upper and lower bounds for the perimeter of the room
   – upper and lower bounds for the area of the room

◎ Assuming that the measurements are correct to the nearest tenth of a metre, calculate
   – upper and lower bounds for the perimeter of the room
   – upper and lower bounds for the area of the room

       Investigate further

Explore

◎ Consider the calculation $\dfrac{a-b}{c-d}$, where a, b, c and d are positive numbers rounded to the nearest whole number

◎ If you want to find the maximum possible value of the answer to this calculation, which of the numbers a, b, c and d should be given their maximum possible value and which of the numbers should be given their minimum possible value? (Try it out with some easy numbers!)

◎ Justify your choice

       Investigate further

Measures

The following exercise tests your understanding of this chapter, with the questions appearing in order of increasing difficulty.

1 a An object has a mass of 12 grams and a volume of 18 cm^3. What is its density?

  b A gold bar has a mass of 30 g and a density of 19.3 g/cm^3. What is its volume?

  c A metal bar measuring 50 cm by 20 cm by 5 cm has a density of 15 g/cm^3. What is the mass of the bar?

  d A roll of copper has a mass of 213.6 g and a volume of 12 cm^3. What is the density of the copper?

  e A boat travels at 6 km/h for 5 h 40 min. How far does it travel?

  f Assume the Sun to be 93 000 000 miles from the Earth, to the nearest million miles. If light travels at 186 000 miles per second, estimate how long it takes for the light from a sudden solar flare to reach the Earth.

2 Write down the lower and upper limits of the following measurements.

  a The Sun is 93 000 000 miles from the Earth to the nearest million miles.

  b Euan's arm is 1 m long to the nearest 10 cm.

  c A parcel weighs 940 g to the nearest 5 g.

  d The 1500 m race was won in a time of 246 s to the nearest second.

  e The number of matches in a box is 100 to the nearest 10 matches.

  f Jane is 35 years old to the nearest year.

  g A stick has a length of 56.4 cm to the nearest mm.

  h A rectangle is said to measure 12.0 cm by 9.6 cm.

   i Calculate the upper and lower bounds for the area of the rectangle if these measurements are both accurate to 1 decimal place.

   ii Calculate the upper and lower bounds for the area if both measurements are accurate to 2 significant figures.

Try a real past exam question to test your knowledge:

3 a The numbers in this calculation are given to 3 significant figures.
   Find the least possible value of $\dfrac{12.3}{15.6 - 7.20}$
   You **must** show all your working.

  b The maximum safe load of a lift is 1500 kg, to the nearest 50 kg.
   The lift is loaded with boxes weighing 141 kg and 150 kg, both weights given to the nearest kilogram.

   Can the lift safely carry 3 boxes weighing 141 kg each and 7 boxes weighing 150 kg each?
   You **must** show all your working.

Spec B, Higher Paper Module 3, Nov 04

5 Real-life graphs

What you should already know ...

- Solve problems involving proportional reasoning such as speed

- $\text{Speed} = \dfrac{\text{distance}}{\text{time}}$

- Plot points of a conversion graph and read off values

- Read from a conversion graph for negative values

- Interpret distance–time graphs

- Convert between miles and kilometres

VOCABULARY

Speed – the rate of change of distance with respect to time. To calculate the average speed, divide the total distance moved by the total time taken. It is usually given in metres per second (m/s) or kilometres per hour (km/h) or miles per hour (mph)

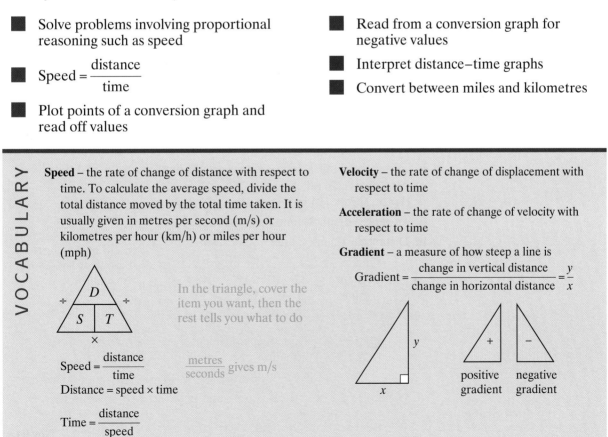

In the triangle, cover the item you want, then the rest tells you what to do

$\text{Speed} = \dfrac{\text{distance}}{\text{time}}$

$\dfrac{\text{metres}}{\text{seconds}}$ gives m/s

$\text{Distance} = \text{speed} \times \text{time}$

$\text{Time} = \dfrac{\text{distance}}{\text{speed}}$

Velocity – the rate of change of displacement with respect to time

Acceleration – the rate of change of velocity with respect to time

Gradient – a measure of how steep a line is

$\text{Gradient} = \dfrac{\text{change in vertical distance}}{\text{change in horizontal distance}} = \dfrac{y}{x}$

positive gradient   negative gradient

Learn 1 Distance–time and velocity-time graphs

Examples:

1 The distance–time graph below shows the journey of a school bus dropping students off at two villages.

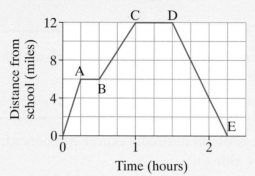

a Describe the journey, giving reasons for the shape of the graph.

b How far are the villages from school?

c During which section is the bus travelling at the fastest speed?

d Calculate the speed of the bus, in miles per hour (mph), during these sections:

   i OA   **ii** CD   **iii** DE

e Calculate the average speed for the whole journey.

a The bus leaves school at a constant speed.

After $\frac{1}{4}$ hour, the bus stops for $\frac{1}{4}$ hour.

It then carries on the journey at a slower speed for $\frac{1}{2}$ hour before stopping again after 12 miles.

It stops for $\frac{1}{2}$ hour and then travels back to school at a constant speed.

b The villages are 6 miles and 12 miles from school.

c OA because the gradient (steepness) of the line is the greatest during this section.

d i Distance = 6 miles, time = $\frac{1}{4}$ hour, so in 1 hour the bus would travel 4×6 miles = 24 miles.

   Hence, speed = 24 miles per hour = 24 mph

ii Distance = 0 miles, time = $\frac{1}{2}$ hour, so speed = 0 mph

iii Distance = 12 miles, time = $\frac{3}{4}$ hour, so:

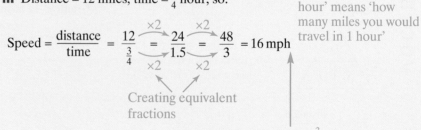

Remember: 'miles per hour' means 'how many miles you would travel in 1 hour'

Creating equivalent fractions

Alternatively, 12 miles in $\frac{3}{4}$ hour is the same as 4 miles in $\frac{1}{4}$ hour or 16 miles in 1 hour

e Total distance = 24 miles, total time = $2\frac{1}{4}$ hours, so

$$\text{Average speed} = \frac{\text{distance}}{\text{time}} = \frac{24}{2\frac{1}{4}} = \frac{48}{4\frac{1}{2}} = \frac{96}{9} = 10.\dot{6} = 11 \text{ mph}$$

2 The graph below describes the performance of an athlete during a race.

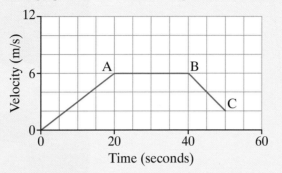

a Describe the race, giving reasons for the shape of the graph.

b How long did it take for the athlete to finish the race?

c Calculate the acceleration, in ms^2, of the athlete during these sections:

    i OA   **ii** AB   **iii** BC

a The athlete accelerates at a constant rate for the first 20 seconds. He then runs at a constant speed for 20 seconds. He then decelerates for the last 10 seconds of the race.

b 50 seconds

c i $\text{Acceleration} = \dfrac{\text{change in velocity}}{\text{change in time}} = \dfrac{6}{20} = 0.3 \text{ m/s}^2$

  ii $\text{Acceleration} = \dfrac{\text{change in velocity}}{\text{change in time}} = \dfrac{0}{20} = 0 \text{ m/s}^2$

The velocity remains at 6 m/s because the athlete is running at a constant speed (neither acceleration nor deceleration)

  iii $\text{Acceleration} = \dfrac{-4}{10} = -0.4 \text{ m/s}^2$

The velocity has decreased from 6 m/s to 2 m/s, that is, the athlete is slowing down (a 'negative acceleration' or 'deceleration')

Apply 1 ⊞

1 Miranda goes out jogging every day. Her journey is shown below:

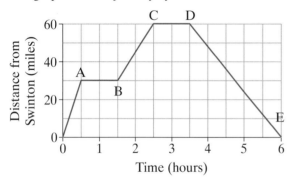

'Miranda runs very fast for the first 2 minutes and then jogs at a constant speed for a short while. She then runs even faster until she is 1500 m from home. She jogs at a constant speed for 4 minutes before running backwards to her home.'

Do you agree with this description of Miranda's journey?

Give reasons for your answer.

2 The graph shows a journey by a car from Swinton.

a Jess works out the speed of the car during section OA as follows:

$$\text{Speed} = \frac{\text{distance}}{\text{time}} = \frac{30}{\frac{1}{2}} = 15 \text{ miles per hour}$$

Do you agree? Give a reason for your answer.

b Jess also calculates the average speed for the whole journey as follows:

$$\text{Average speed} = \frac{\text{total distance}}{\text{total time}} = \frac{60}{6} = 10 \text{ miles per hour}$$

Do you agree? Give a reason for your answer.

3 Jim the courier is delivering parcels from his depot in Nelson.
He leaves the depot at 9.45 a.m.

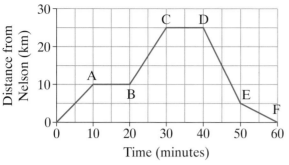

a How many times does Jim stop to deliver parcels?

b At what time does Jim arrive at his first delivery?

c How far does Jim travel in the first 10 minutes?

d Calculate Jim's speed during section BC in kilometres per hour (km/h).

e Calculate Jim's average speed over the first half hour of the journey.

f At what time does Jim start to return to the depot?

g During which section is Jim travelling the fastest?

h Calculate Jim's average speed for the whole journey.

i Did Jim break the motorway speed limit of 70 mph during any part of his journey?

4 Two snails, Sandra and Brian, are racing along a 30 cm ruler.
Write a commentary for the race.
You should include key distances, times and speeds in your commentary.

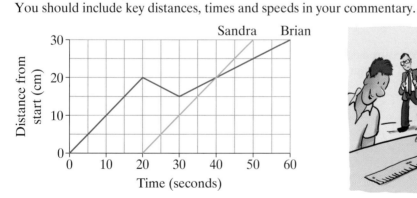

5 Pat walks to her local park for 15 minutes at 4 km/h. She sits on a bench for 10 minutes and then walks back home at a speed of 3 km/h. Draw a distance–time graph representing Pat's journey.

6 Bill and Ben both leave their homes at 10.30 a.m.
Bill is travelling the 155 miles from his home in Cheltenham to Manchester.
He travels along the M5 at 60 mph for 50 minutes and then is
held up in a traffic jam for half an hour as the M5 joins the M6.
He then travels the remainder of the journey at 70 mph.
Ben leaves Manchester at the same time that Bill leaves Cheltenham.
He travels to Cheltenham at a constant speed of 50 mph.

a Draw distance–time graphs showing Bill's and Ben's journeys.

b At what times do Bill and Ben arrive at their destinations?

c At what time are Bill and Ben the same distance from Cheltenham?

d Calculate Bill's average speed for the whole journey.

7 Describe the 50 km race between two cyclists shown in the graph.
You should include key distances, times and speeds in your commentary.

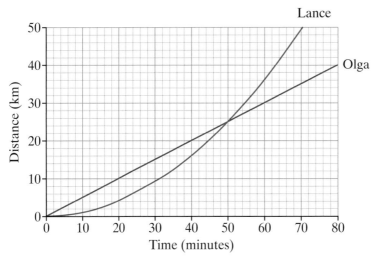

8 The speed of a train is measured as shown below.
   a Describe the journey of the train.

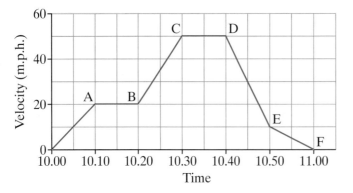

b During which section is the train travelling the fastest?

c Calculate the acceleration of the train at:

   i 10.05   **ii** 10.15   **iii** 10.45

9

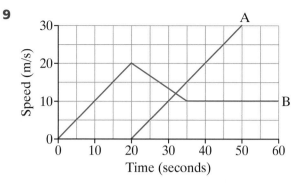

'Person A travels at a constant speed for the whole journey.'
'Person B travels at a constant speed for 20 seconds then turns around and travels in the opposite direction for 15 seconds. Person B then remains stationary for 25 seconds.'
Do you agree with the two descriptions? Give reasons for your answers.

Learn 2 Other real-life graphs: linear and non-linear

Example: The graph shows how the price of a house has changed over the last eight years.

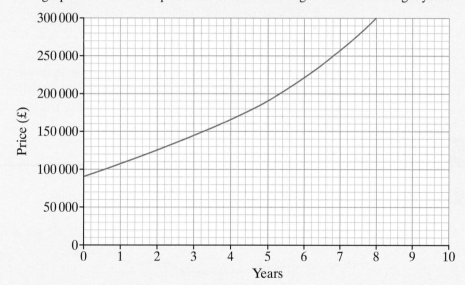

a Explain the shape of the graph.

b How much was the house originally bought for?

c When had the price of the house increased by 50%?

d Would it be appropriate to use the graph in 10 years' time to estimate the price of the house?

a The price of the house is increasing at an increasing rate.

Possible reasons might include that it is located in an area of high demand such as commuter belt, area of outstanding beauty, catchment area of a highly successful school, etc.

b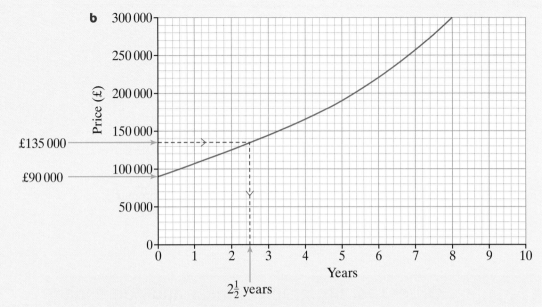

$2\frac{1}{2}$ years

The house was originally bought for £90 000.

c 50% increase on £90 000 = £135 000.
After approx $2\frac{1}{2}$ years.

The number of years can be read off from the graph

d No, it would not be appropriate.
Results are only valid in the range of the data that has been plottted.
The graph would give a value of £1.8 million after 18 years!

Apply 2

1

Mortgage Interest Rates (%) in 2004–2005

a Using the graph, copy and complete the following table:

2004											
	J	**F**	**M**	**A**	**M**	**J**	**J**	**A**	**S**	**O**	**N**	**D**
%	5.75					5.65			5.5		

2005											
	J	**F**	**M**	**A**	**M**	**J**	**J**	**A**	**S**	**O**	**N**	**D**
%												6.75

b What do you notice about the mortgage interest rates?

c During 2004 and 2005, Andrew decided to have a fixed mortgage at 5.8% rather than a variable one. Do you think he made a good decision? Give a reason for your answer.

2 Asheed's parents are trying to decide which mobile phone deal to buy – Purple or C4.

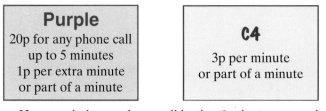

Purple
20p for any phone call
up to 5 minutes
1p per extra minute
or part of a minute

C4
3p per minute
or part of a minute

a How much does a phone call lasting 6 minutes cost using:

i Purple **ii** C4?

b How much does a phone call lasting 10 minutes cost using:

i Purple **ii** C4?

c Copy the axes below. Plot two graphs showing how the cost of a phone call (pence) varies with the length of the phone call (minutes) for each mobile deal.

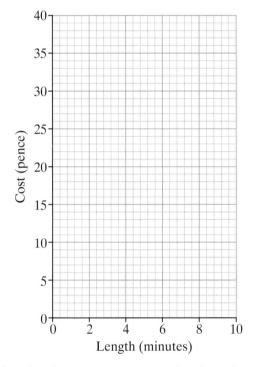

d Asheed's parents are buying the phone for Asheed to make short phone calls to let them know where she is and when she needs a lift home.

Which phone would you recommend the parents buy?
Give a reason for your answer.

3 a Each of these containers is filled with water flowing from a tap at a steady rate.

Match each container to the correct graph showing how the depth (*d*) of water varies with time (*t*).

b Draw container **C** and graph **ii**.

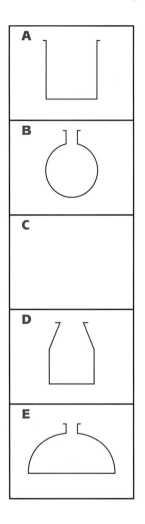

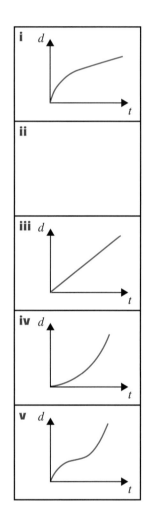

7 Alfredo, the Great Human Cannonball, is fired from Big Bertha.

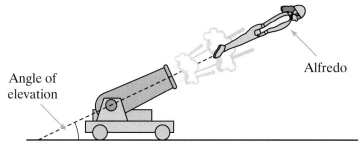

Angle of
elevation

Alfredo

The paths of Alfredo fired from Big Bertha at different angles of elevation
are shown below.

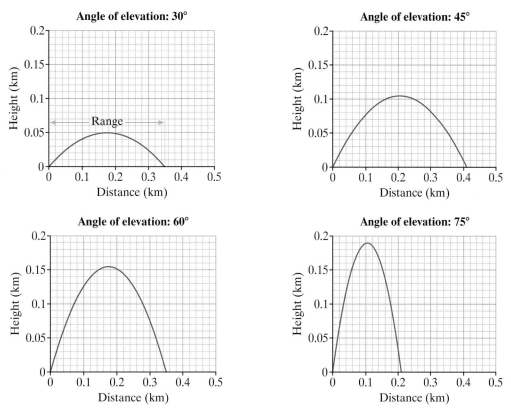

a Which angle of elevation gives the greatest range?
Write down the range.

b Which angle of elevation fires Alfredo to the greatest height?
Write down the height.

c Using the information in the diagram below, estimate a suitable range
of angles that will enable Alfredo to fly safely from Big Bertha to the
landing net.

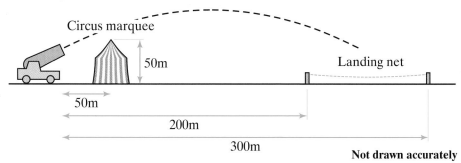

Circus marquee

50m

Landing net

50m

200m

300m

Not drawn accurately

63

8 The tidal pattern of 'The Masnahs' over a two-day period is shown in this graph.

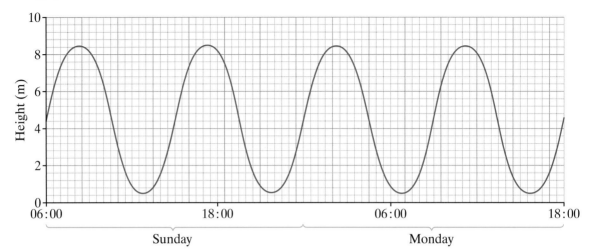

a At what times is high tide on:

   i Sunday   **ii** Monday?

b At what times is low tide on:

   i Sunday   **ii** Monday?

c A great time to surf is when the tide is coming in. At what times is this possible during the two days?

d Andy wants to go sea kayaking but decides to avoid low tide as he doesn't want to carry the kayak down the beach (minimum tide height for kayaking = 6 m). He only wants to kayak when the tide is going out. At what times should he book the kayak?

e The best Sunday lunch is served in Lower Masnah. Is it possible to walk along the beach from Upper Masnah? (The headland is only passable if the tide is below 3 m.)

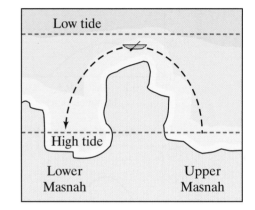

Explore

A slug moves forward every day

Unfortunately, due to depleting energy levels, every day it only moves half of the previous day's distance

The slug moves 1 m on Day 1

◎ How far does the slug move on Day 2?

◎ What is the total distance travelled after two days?

◎ What is the total distance travelled after   **a** 3 days   **b** 5 days   **c** 10 days?

◎ Plot these points on a graph

> Investigate further

Explore

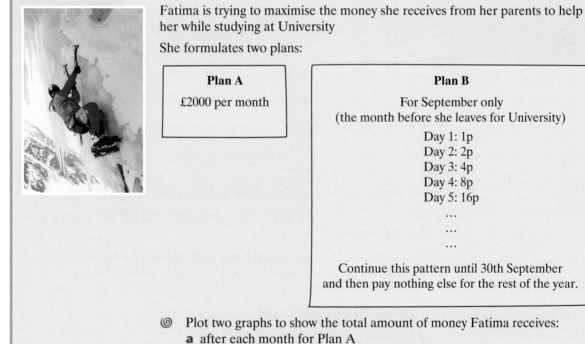

Fatima is trying to maximise the money she receives from her parents to help her while studying at University

She formulates two plans:

Plan A
£2000 per month

Plan B

For September only
(the month before she leaves for University)

Day 1: 1p
Day 2: 2p
Day 3: 4p
Day 4: 8p
Day 5: 16p
...
...
...

Continue this pattern until 30th September
and then pay nothing else for the rest of the year.

◎ Plot two graphs to show the total amount of money Fatima receives:
  a after each month for Plan A
  b after each day for Plan B

> Investigate further

Real-life graphs

The following exercise tests your understanding of this chapter, with the questions appearing in order of increasing difficulty.

1 Bill walks his dog Hugo. The distance they are from home is shown in the graph below.

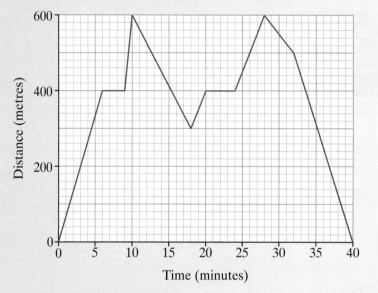

a After how long are they first:

i 200 metres from home

ii 550 metres from home?

b Bill stands and talks to a friend for 4 minutes. When did this happen?

c How long is it from the first to the second time that Bill and Hugo are 600 m from home?

d At one stage it began to rain. Bill and Hugo headed for home. Fortunately the shower soon stopped and Bill decided to resume their walk.

i When did the shower start?

ii How long did the shower last?

iii How far did they walk towards home?

e How far did Bill and Hugo walk altogether?

f Use the distance–time graph to calculate the speed at which Bill and Hugo are walking during these times (give your answers in metres per minute):

i the first 6 minutes

ii from 10 to 18 minutes

iii from 24 to 28 minutes

iv the last 8 minutes.

2 The graph shows the profit, £y, made from making x wine glasses.

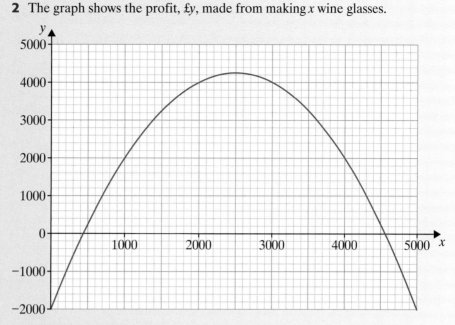

Use the graph to find:

a how many wine glasses the firm needs to produce if it is to make the maximum profit

b the minimum number of wine glasses the firm needs to produce if it is not to make a loss

c the range of values of x for which the profit is more than £3250.

3 The velocity–time graph for a train, moving between two stations, is shown below.

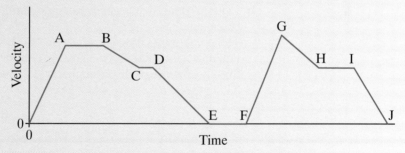

Comment on the movement of the train during each of the lettered sections OA, AB, BC, etc.

Try a real past exam question to test your knowledge:

4 a Liquid is poured at a steady rate into the bottle shown in the diagram.

As the bottle is filled, the height, h, of the liquid in the bottle changes.

Which of the five graphs below shows this change? Give a reason for your choice.

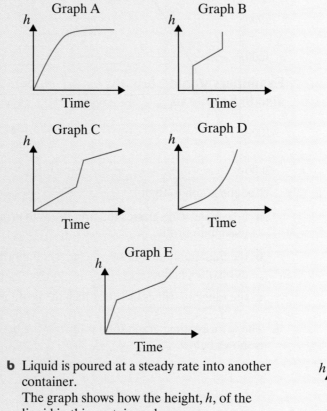

b Liquid is poured at a steady rate into another container.

The graph shows how the height, h, of the liquid in this container changes.

Sketch a picture of this container.

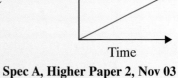

Spec A, Higher Paper 2, Nov 03

OBJECTIVES

D **Examiners would normally expect students who get a D grade to be able to:**

Substitute numbers into more complicated formulae such as $C = \dfrac{(A + 1)D}{9}$

C **Examiners would normally expect students who get a C grade also to be able to:**

Find a solution to a problem by forming an equation and solving it

Rearrange linear formulae such as $p = 3q + 5$

B **Examiners would normally expect students who get a B grade also to be able to:**

Rearrange formulae that include brackets, fractions and square roots

A **Examiners would normally expect students who get an A grade also to be able to:**

Rearrange formulae where the variable appears twice

What you should already know ...

- The four rules applied to negative numbers
- Calculate the squares, cubes and other powers of numbers
- Write formulae using letters and symbols
- Simplify expressions by collecting like terms

- Solve linear equations
- Expand brackets such as $3(x + 4)$
- The four rules applied to fractions
- Factorise an expression such as $2x + 6$ into $2(x + 3)$
- Round numbers to a given number of significant figures

VOCABULARY

Expression – a mathematical statement written in symbols, for example, $3x + 1$ or $x^2 + 2x$

Equation – a statement showing that two expressions are equal, for example, $2y - 7 = 15$

Formula – an equation showing the relationship between two or more variables, for example, $E = mc^2$

Identity – two expressions linked by the $\equiv$ sign are true for all values of the variable, for example, $3x + 3 \equiv 3(x + 1)$

Variable – a symbol representing a quantity that can take different values such as x, y or z

Term – a number, variable or the product of a number and a variable(s) such as $3, x$ or $3x$

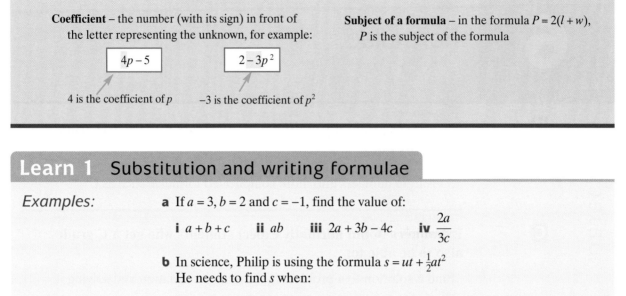

Coefficient – the number (with its sign) in front of the letter representing the unknown, for example:

$$4p - 5$$

4 is the coefficient of p

$$2 - 3p^2$$

−3 is the coefficient of p^2

Subject of a formula – in the formula $P = 2(l + w)$, P is the subject of the formula

Learn 1   Substitution and writing formulae

Examples:

a If $a = 3$, $b = 2$ and $c = -1$, find the value of:

　i $a + b + c$　**ii** ab　**iii** $2a + 3b - 4c$　**iv** $\dfrac{2a}{3c}$

b In science, Philip is using the formula $s = ut + \frac{1}{2}at^2$
He needs to find s when:

　i $u = 4$, $t = 3$ and $a = 5$　**ii** $u = 3$, $t = 4$ and $a = 0.5$

c I think of a number, multiply it by 6 and add 1. The answer is 43.
What was the number?

This part of the question tells you that you need to write an equation

d Andrew needs to work out the sizes of the angles in this diagram.
Form an equation to help him work out the answer.
What size are the three angles?

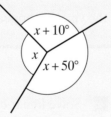

$x + 10°$

x

$x + 50°$

Not drawn accurately

a i　$a + b + c$
　　$= 3 + 2 + -1$　——— Replace the letters with the values given
　　$= 4$

ii　ab　←——— Remember that ab means $a \times b$
　　$= 3 \times 2$
　　$= 6$

iii　$2a - 3b + 4c$
　　$= (2 \times 3) - (3 \times 2) + (4 \times -1)$
　　$= 6 - 6 - 4$
　　$= -4$　　$+ -4 = -4$

Remember to show all stages of your working; you gain method marks for this in an examination

iv　$\dfrac{2a}{3b}$

　　$= \dfrac{(2 \times 3)}{(3 \times 2)}$

　　$= \dfrac{6}{6}$

　　$= 1$

b i $s = ut + \frac{1}{2}at^2$

$s = (4 \times 3) + (\frac{1}{2} \times 5 \times 3^2)$

$s = 12 + 22.5$

$s = 34.5$

First, write down the formula to use; next, substitute the given values into the formula

$\frac{1}{2}at^2$ means $\frac{1}{2} \times a \times t^2$ so only the t is squared

ii $s = ut + \frac{1}{2}at^2$

$s = (3 \times 4) + (\frac{1}{2} \times 0.5 \times 4^2)$

$s = 12 + 4$

$s = 16$

c First, write an equation:

Let the unknown number be	x
Multiply by 6	$6x$
Add 1	$6x + 1$
The answer is 43	$6x + 1 = 43$ ⟵——— Create an equation

Next, solve the equation:

$6x + 1 - 1 = 43 - 1$ ⟵——— Subtract 1 from both sides

$6x = 42$

$6x \div 6 = 42 \div 6$ ⟵——— Divide both sides by 6

$x = 7$

d $x + x + 10° + x + 50° = 360°$ ⟵——— Use the fact that angles at a point add up to 360° to help you form an equation

$3x + 60° = 360°$ ⟵——— Simplify by adding like terms together

Solve the equation:

$3x + 60° - 60° = 360° - 60°$ ⟵——— Subtract 60 from both sides

$3x = 300°$

$3x \div 3 = 300° \div 3$ ⟵——— Divide both sides by 3

$x = 100°$

The three angles are $x = 100°$, $x + 10° = 110°$ and $x + 50° = 150°$

Check to see that your answers add up to 360°

Apply 1

1 Find the value of each of these expressions when $x = 3$:

   a $4x$ **b** $3x - 2$ **c** x^2 **d** $x^2 + 4x - 6$

2 Find the value of each of these expressions when $y = -2$:

   a $3y$ **b** $4y + 6$ **c** $y^2 - 3$ **d** $2y^2 + 3y$

3 If $x = 2$, $y = 3$, $s = 4$ and $t = 0.5$, find the value of:

a $2x$

b $6y$

c $3y + x$

d $3s - t$

e $\dfrac{s}{2} + 1$

f $\dfrac{4s}{x}$

g xy

h $2s - t + y$

i xys

j $st \div x$

k $xy - st$

l $\dfrac{st}{xy}$

m $t(s + x)^2$

n $y^2(x - t)$

o $\dfrac{sx^2y}{t}$

p $\dfrac{tx^2y}{s^2}$

4 If $p = 3$, $q = 6$ and $r = -7$, find the value of:

a $q + 3$

b $12 - r$

c $3r$

d pqr

e $\dfrac{12}{p}$

f $p + 3q + 4r$

g $\dfrac{pq}{9}$

h $p + 4q - 2r$

i $\dfrac{r}{7}$

j $pr - qp$

k $r + 3pq$

l $\dfrac{q}{r}$

m $p(r + q)$

n $\dfrac{p^2r}{(p + q)}$

o $q^2(r - p)$

p $q(p - r)$

5 If $a = 2$, $b = -3$ and $c = 5$, find the value of:

a a^2

b b^2

c abc

d $a^2 + b$

e $b^2 - c$

f $3c^2$

g $2c - 4a$

h a^2c

i abc^2

j $a^2 + b^2$

k $b^2 - a^3$

l $a^4 - b^3$

m $c^2(b - a)$

n $\sqrt{\dfrac{a^2c^3}{(a - b)}}$

o $\dfrac{(a - b)^2}{c}$

p $a(c - b)^2$

6 Emily has worked out the answer to question **5f** as 225. Is she correct? Give a reason for your answer.

7 Antony is using the formula $C = \dfrac{(A + 1)D}{9}$

He knows that $D = 30$ and $A = -7$
Work out the value of C.

8 The formula for finding the perimeter of a rectangle is $P = 2l + 2w$, where P represents perimeter, l represents length and w represents width. Find the perimeter of each of the following rectangles:

a Length = 12 cm; Width = 5 cm

b Length = 52 cm; Width = 22 cm

c Length = 3.5 cm; Width = 2.5 cm

9 Get Real!

The cost in pounds (C) of framing a picture depends on its length in cm (l) and its width in cm (w).

The formula used by Framing For You is $C = 3l + 2w$

Find the cost of framing a picture that is:

a 50 cm long and 20 cm wide

b 30 cm long and 20 cm wide

c 1.2 m long and 60 cm wide.

10 Get Real!

Tasty Pizzas works out its delivery charge with the formula:

Delivery charge = number of pizzas × 50p + £1.25

a Find the delivery charge for three pizzas.

b Richard pays a delivery charge of £4.25. How many pizzas did he order?

11 Get Real!

Mrs Sturgess is working out the wages at the factory. She uses the formula:

Wages equal hours worked multiplied by rate per hour.

a Write this formula in algebra.

b How much does Eileen earn if she is paid £7.50 an hour and she works for 35 hours?

c How much does Freddy earn if he works for 25 hours and is paid £8.25 an hour?

d How long did Louise work if she earned £465.50? Her hourly rate is £12.25

12 Get Real!

In science, Andrew is using the formula $V = IR$

Find V if:

a $I = 3$ and $R = 52$ **b** $I = 2.5$ and $R = 63$

13 Get Real!

In maths, Sarah is using the formula $A = \pi r^2$

Find A if:

a $\pi = 3.14$ and $r = 15.5$ **b** $\pi = 3.14$ and $r = 8$

14 Get Real!

In geography, Keith is using the formula $C = \dfrac{5(F - 32)}{9}$

Find C if:

a $F = 68$ **b** $F = 32$ **c** $F = -31$ **d** $F = -13$

15 Get Real!

In science, Briony is using the formula $s = ut + \frac{1}{2}at^2$

Find s if:

a $u = 3, t = 6.5$ and $a = 14.3$ **b** $u = 4, t = 3$ and $a = 8$

16 Get Real!

In maths, Didier is using the formula $V = \dfrac{4\pi r^3}{3}$

Find V if:

a $\pi = 3.14$ and $r = 12$ **b** $\pi = 3.14$ and $r = 2.5$

17 Get Real!

In maths, Emily is using the formula $V = \frac{1}{3}\pi r^2 h$ to work out the volume of a cone.

a Find V if $\pi = 3.14$, $r = 3$ cm and $h = 2.5$ cm

b A cone has a volume of 185 cm^3.
Calculate its height (h) if $r = 5$ cm and $\pi = 3.14$
Give your answer correct to the nearest millimetre.

18 Get Real!

Tom is calculating the interest on some savings.

The formula he uses is $I = \dfrac{PTR}{100}$

a If the principal (P) = £300, the time (T) = 2 years and the rate (R) = 2.5%, calculate the interest.

b How long would he have to save if he wanted £95 interest?

19 Write an equation and solve it to find the unknown number in each of these questions.
Let the unknown number be z.

a Think of a number. Add 3. The answer is 29.

b Think of a number. Multiply it by 3. Add 5. The answer is 17.

c Think of a number. Multiply it by 4. Subtract 5. The answer is 23.

d Think of a number. Add 3. Multiply by 4. The answer is 12.

e Think of a number. Subtract 7. Multiply by 2. The answer is 6.

20 Get Real!

A cookery book gives this rule to roast a chicken:

> Allow 20 minutes per pound plus 20 minutes.

a Write a formula for this rule. Use c to represent the time needed to roast a chicken and p to represent the number of pounds.

b How long would a 3-pound chicken take to roast?

21 Write down an equation for each of these diagrams.
Solve it to find the value of *x*.

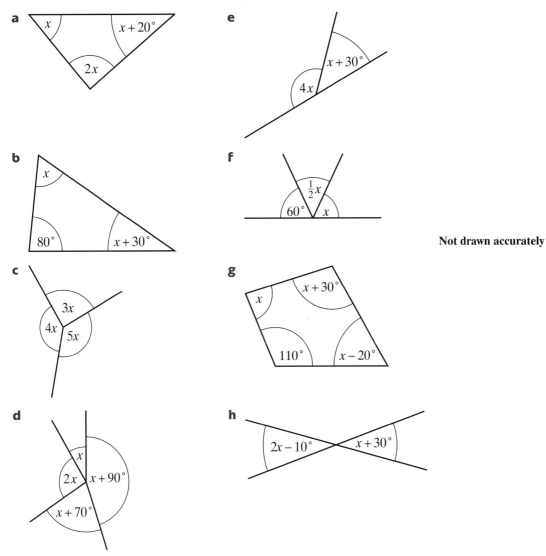

Not drawn accurately

22 The length of a rectangle is 5 cm more than its width. The width is *b* cm.

  a Write down an expression for the length.

The perimeter of the rectangle is 90 cm.

  b Find the area of the rectangle.

23 A square has sides of length *x* + 2. The perimeter is 32 cm.
Find the area of the square.

24 Katie goes on holiday with $*x*. Sam has three times as much as Katie.
James has $5 more than Katie. Altogether they have $250.
How much do they each have?

Explore

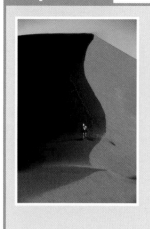

Andy's Uncle Jim is 70 today and he has decided to give Andy some money
Andy can choose one of four options:

◎ **Option 1:** £100 now, £90 next year, £80 the year after and so on

◎ **Option 2:** £10 now, £20 next year, £30 the year after and so on

◎ **Option 3:** £10 now, 1.5 times as much next year, 1.5 times as much the year
after and so on

◎ **Option 4:** £1 now, £2 next year, £4 the year after, £8 the year after and so on

Each of the options will only operate whilst Uncle Jim is alive

Which option should Andy choose?

(Investigate further)

Learn 2 Changing the subject of a formula

Examples:

a Make x the subject of the formula $y = 35x + 350$

b Make r the subject of the formula $B = cr^2$

c Make x the subject of these formulae:

i $a = \dfrac{b}{x}$ **ii** $f(f - x) = g$ **iii** $h = \sqrt{\dfrac{j}{x}}$ **iv** $kx + l = mx + n$

a
$$y = 35x + 350$$
$$y - 350 = 35x + 350 - 350 \qquad \text{Subtract 350 from both sides}$$
$$y - 350 = 35x$$
$$\frac{y - 350}{35} = \frac{35x}{35} \qquad \text{Divide both sides by 35}$$
$$\frac{y - 350}{35} = x \qquad \text{x is now the subject of the formula}$$
$$x = \frac{y - 350}{35} \qquad \text{Put x on left-hand side as the subject of the formula}$$

b
$$B = cr^2$$
$$\frac{B}{c} = \frac{cr^2}{c} \qquad \text{Divide both sides by c}$$
$$\frac{B}{c} = r^2$$
$$\sqrt{\frac{B}{c}} = \sqrt{r^2} \qquad \text{Find the square root of both sides of the equation}$$
$$\pm\sqrt{\frac{B}{c}} = r \qquad \text{Remember that square root can be positive or negative}$$
$$r = \sqrt{\frac{B}{c}}$$

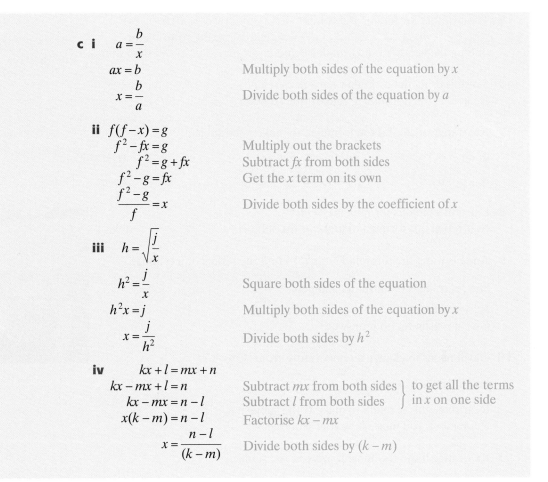

c i $a = \dfrac{b}{x}$

$ax = b$ Multiply both sides of the equation by x

$x = \dfrac{b}{a}$ Divide both sides of the equation by a

ii $f(f - x) = g$

$f^2 - fx = g$ Multiply out the brackets

$f^2 = g + fx$ Subtract fx from both sides

$f^2 - g = fx$ Get the x term on its own

$\dfrac{f^2 - g}{f} = x$ Divide both sides by the coefficient of x

iii $h = \sqrt{\dfrac{j}{x}}$

$h^2 = \dfrac{j}{x}$ Square both sides of the equation

$h^2 x = j$ Multiply both sides of the equation by x

$x = \dfrac{j}{h^2}$ Divide both sides by h^2

iv $kx + l = mx + n$

$kx - mx + l = n$ Subtract mx from both sides ⎫ to get all the terms

$kx - mx = n - l$ Subtract l from both sides ⎬ in x on one side

$x(k - m) = n - l$ Factorise $kx - mx$

$x = \dfrac{n - l}{(k - m)}$ Divide both sides by $(k - m)$

Apply 2

1 Rearrange the formula $M = l + 32$ to make l the subject.

2 Rearrange each of these formulae to make x the subject:

a $y = x - 22$ **f** $ax - b^2 = c^2$ **k** $kx - l = m$

b $x + b = c$ **g** $d = 7x - 50$ **l** $nx + p = q$

c $y = bx$ **h** $ex = f + g$ **m** $r + sx = t$

d $a + x = c$ **i** $3x = h - i$ **n** $u + vx = 2w$

e $gx + p^2 = s^2$ **j** $xy = j$ **o** $y - x = z$

3 The formula for finding the circumference of a circle is $C = \pi d$
Rearrange this formula to make d the subject.

4 The formula for finding the speed of an object is $s = \dfrac{d}{t}$ where s stands for
speed, d for distance and t for time. Rearrange the formula to make d
the subject.

5 The formula for finding the volume of a cylinder is $V = \pi r^2 h$
where V = volume, r = radius and h = height. Write the formula for
finding the height of the cylinder if you know the volume and the radius.

6 Sarah is using the formula $v = u + at$ in science.
Rearrange the formula to make a the subject.

7 Rearrange each of these formulae to make y the subject:

a $ay^2 = b$ **c** $y^2 z = (e + f)$ **e** $k^2 + y^2 = l^2$

b $y^2 c = d$ **d** $gy^2 - h = j$ **f** $m^2 - y^2 = n^2$

8 Einstein's famous formula is $E = mc^2$
Rearrange this formula to make m the subject.

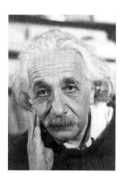

9 Anne is using the formula $V = \pi r^2 h$ to find the radius of a cylinder.

She rearranges the formula and gets $r = \dfrac{\sqrt{V}}{\pi h}$

Is she correct?
Give a reason for your answer.

10 Gavin needs to change a temperature from °C to °F.

He finds this formula: $F = \dfrac{9C}{5} + 32$

Rearrange this formula to make C the subject.
Remember, you must show your working.

11 Rearrange these formulae to make x the subject.

a $p = \dfrac{x}{q}$ **d** $z = a(b + x)$ **g** $\sqrt{x} = m$ **i** $e = 2x + xf$

b $\dfrac{a}{b} = \dfrac{x}{c}$ **e** $n - x = p$ **h** $t + u = v\sqrt{x}$ **j** $xq^2 = r(x + s)$

c $l(x + m) = n$ **f** $y = a(b - x)$

In questions **12** to **15** give the answer to 3 s.f. where necessary. Use $\pi = 3.14$

12 The formula to find the length of the diagonal (d) of a rectangle l units
long and w units wide is $d = \sqrt{l^2 + w^2}$

    a Make l the subject of the formula.

    b Find the length of l when $d = 25$ cm and $w = 15$ cm

13 a Find a formula for the sum (S) of three consecutive whole numbers n,
$n + 1$, and $n + 2$.

    b Rearrange the formula to make n the subject.

    c Find the three consecutive whole numbers whose sum is 366.

14 a Rearrange the formula $v^2 = u^2 + 2as$ to make u the subject and hence
find the value of u when $v = 10$, $a = 2$ and $s = 9$

    b Rearrange the formula to make a the subject, hence find the value of
a when $s = 8$, $u = 7$ and $v = 9$

15 Rearrange the formula $t = 2\pi\sqrt{\dfrac{I}{mgh}}$ to make I the subject

Hence find the value of I when $m = 2$, $h = \frac{1}{2}$ and $t = 18$

Take $g = 10$

16 Rearrange the formula $y = \dfrac{x-a}{x-b}$ to make x the subject.

17 Emily and Anthony are rearranging the formula $T = 2\pi\sqrt{\dfrac{l}{g}}$ to make l the

subject. Emily gets $\dfrac{T^2 g}{2\pi} = l$ for her answer and Anthony gets $\dfrac{T^2 g}{4\pi} = l$

Is either of them correct? Give a reason for your answer.

18 Make the expression in brackets the subject of the formula.

a $x(a - b) = a(b - x)$ (x) **c** $E = \dfrac{1}{2}m(v^2 - u^2)$ (u) **e** $y = \dfrac{a+x}{x-3}$ (x)

b $\dfrac{a}{\sin A} = \dfrac{b}{\sin B}$ $(\sin A)$ **d** $\dfrac{x+b}{c} = \dfrac{x+c}{b}$ (x) **f** $R = \dfrac{ab}{a+b}$ (b)

Formulae

ASSESS

The following exercise tests your understanding of this chapter, with the questions appearing in order of increasing difficulty.

1 Niamh measures the depth of a well, D m, by dropping a stone down it and timing the splash. The formula she uses is $D = 5t^2$, where t is in seconds. Use this formula to find:

    a The depth of the well if she hears the splash after:

      i 3 seconds   **ii** 12 seconds.

    Another well is 180 m deep.

    b How long will it take to hear the splash if Niamh drops a stone down this well?

2 Use the formula $d = \dfrac{x+4}{10}$ to find:

    a the value of d when $x = 3$

    b the value of d when $x = 48$

    c the value of x when $d = 10$

3 a Make y the subject of the straight line equation $4x + 7y = 8$

b The change in momentum, M, of a body of mass m when it moves from a speed u to a new speed v is given by the formula $M = mv - mu$
Make v the subject.

4 The period of swing, in seconds, of a simple pendulum is given by the formula $T = 2\pi\sqrt{\dfrac{L}{10}}$, where L is the length of the pendulum.

a Make L the subject of the formula.

b A pendulum has a swing of 3 seconds. Find its length. (Take $\pi = 3.14$)

5 Make x the subject of these equations.

a $6(x + t) = 2(x + 7)$ **b** $\dfrac{x - 5}{p} = \dfrac{x + 4}{q}$ **c** $\dfrac{(x - 5)}{(x + 5)} = w$

Try a real past exam question to test your knowledge:

6 A shape is made from two trapeziums.

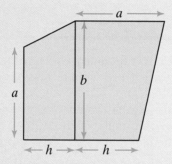

Not drawn accurately

The area of this shape is given by:

$$A = \dfrac{h}{2}(a + b) + \dfrac{b}{2}(a + h)$$

Rearrange the formula to make a the subject.

Spec A, Higher Paper 2, Nov 05

7 Construction

OBJECTIVES

D ▸ **Examiners would normally expect students who get a D grade to be able to:**

Draw a quadrilateral such as a kite or a parallelogram with given measurements

Understand that giving the lengths of two sides and a non-included angle may not produce a unique triangle

Construct and recognise the nets of 3-D solids such as pyramids and triangular prisms

Draw plans and elevations of 3-D solids

C ▸ **Examiners would normally expect students who get a C grade also to be able to:**

Construct the perpendicular bisector of a line

Construct the perpendicular from a point to a line

Construct the perpendicular from a point on a line

Construct angles of 60° and 90°

Construct the bisector of an angle

What you should already know ...

- Use a protractor and a pair of compasses
- Different types of angles and triangles
- Use bearings

- Round decimals to the nearest whole number
- Use a scale on a map and scale drawing

VOCABULARY

Bearing – an angle measured clockwise from North; all bearings should be written as three figure numbers, for example, 125° or 045°

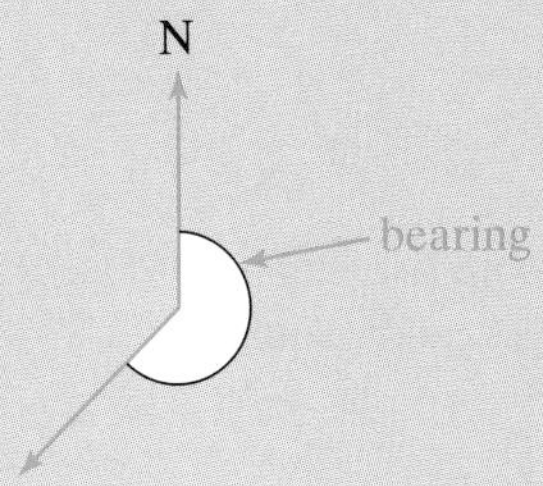

N

bearing

Equilateral triangle – a triangle with 3 equal sides and 3 equal angles – each angle is 60°

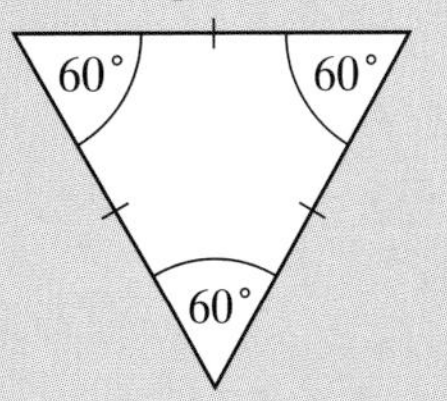

60°   60°

60°

Isosceles triangle – a triangle with 2 equal sides and 2 equal angles; the equal angles are called **base angles**

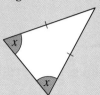

Right-angled triangle – a triangle with one angle of 90°

Congruent – exactly the same size and shape; one of the shapes might be rotated or flipped over

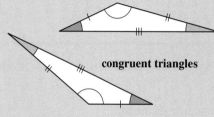

congruent triangles

Perpendicular lines – two lines at right angles to each other

Parallel lines – two lines that never meet and are always the same distance apart

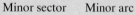

Arc (of a circle) – part of the circumference of a circle; a minor arc is less than half the circumference and a major arc is greater than half the circumference

Minor sector   Minor arc

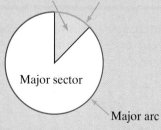

Major sector

Major arc

Bisect – to divide into two equal parts

Midpoint – the middle point of a line

Perpendicular bisector – a line at right angles to a given line that also divides the given line into two equal parts

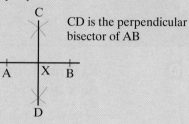

CD is the perpendicular bisector of AB

Angle bisector – a line that divides an angle into two equal parts

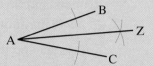

AZ is the angle bisector of angle BAC

Regular polygon – a polygon with all sides and all angles equal

Kite – a quadrilateral with two pairs of equal adjacent sides

Trapezium (pl. **trapezia**) – a quadrilateral with one pair of parallel sides

Parallelogram – a quadrilateral with opposite sides equal and parallel

Rhombus – a quadrilateral with four equal sides and opposite sides parallel

Solid – a three-dimensional shape

Cube – a solid with six identical square faces

Cuboid – a solid with six rectangular faces (two or four of the faces can be squares)

Face – one of the flat surfaces of a solid

Vertex (pl. **vertices**) – the point where two or more edges meet

Edge – a line segment that joins two vertices of a solid

Prism – a three-dimensional solid with two cross-sectional faces that are identical polygons, parallel to each other; all other faces are either parallelograms or rectangles

Prisms are named according to the cross-sectional face; for example,

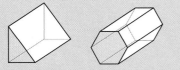

Triangular prism   Hexagonal prism   Parallelogram prism

Cylinder – a prism with a circle as a cross-sectional face

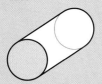

Pyramid – a solid with a polygon as the base and one other vertex; all the vertices of the base are joined to this vertex forming triangular faces. Pyramids are named according to their base, for example,

Square pyramid   Triangular pyramid

Regular tetrahedron – a triangular pyramid with equilateral triangles as its sides

Cone – a pyramid with a circular base and a curved surface rising to a vertex

Hemisphere – a half sphere

Net – a two-dimensional shape made of polygons that can be folded to make a three-dimensional solid, for example,

Net of a cuboid   Net of a triangular prism

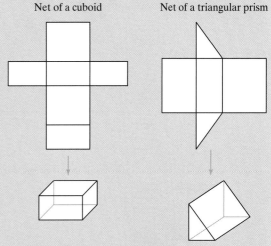

Plan – a diagram of a 3-D solid showing the view from above; these diagrams show a square-based pyramid and its plan

In this drawing the top of the pyramid is vertically above the centre of the base

Plan

Plan

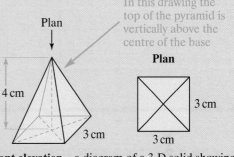

Front elevation – a diagram of a 3-D solid showing the view from the front, for example,

Front elevation

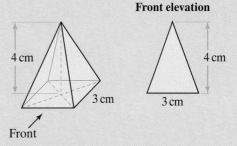

Front

In some cases, as in this prism, the elevation from the front is the same as its cross-section

They are congruent trapezia

Cross-section

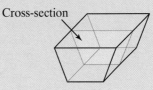

Side elevation – a diagram of a 3-D solid showing the view from the side; sometimes the elevation from the side of the shape is the same as the front elevation, for example,

Unseen edges are shown as dotted lines, for example, the side elevation of this cylindrical container has three unseen edges. (The front elevation is the same)

Front/side elevation

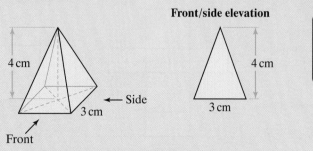

4 cm

← Side

3 cm

4 cm

3 cm

↑
Front

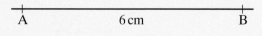 ← S

Usually, however, they will be different, for example,

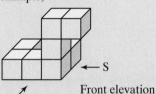 ← S

↑
F

Front elevation viewed from F

Side elevation viewed from S

Learn 1   Drawing 2-D shapes

Examples: **a**   Draw a triangle ABC with AB = 6 cm, angle A = 47° and angle B = 32°.

Step 1   Draw a sketch. You don't need to use a ruler, just draw your sketch freehand.

'Draw' means you can use any piece of equipment (for example, a protractor, a ruler) to produce an accurate diagram

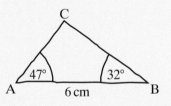

C

47°   32°

A   6 cm   B

Step 2   Draw your first line a little longer than 6 cm, then mark off A and B. This will give you a cross on which to put the centre of your protractor.

A   6 cm   B

Step 3 Use your protractor to draw an angle of 47° at A.

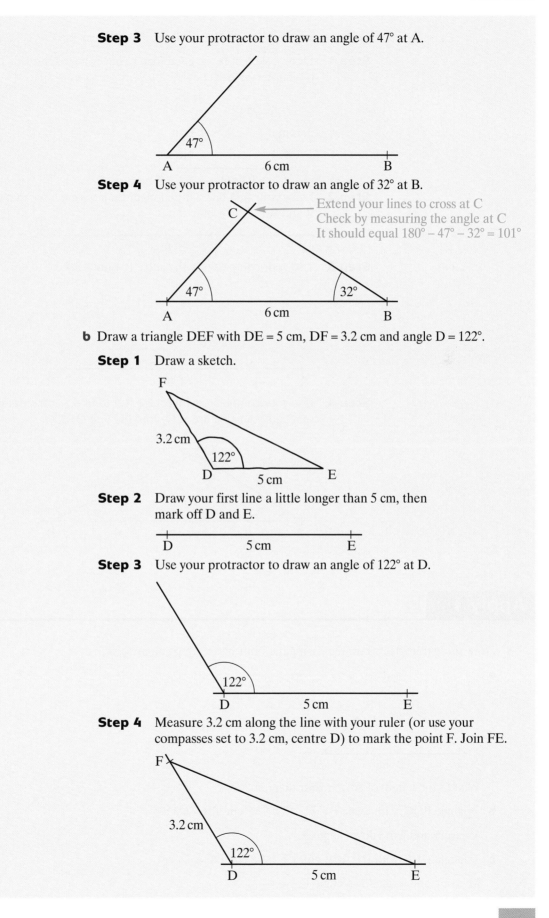

Step 4 Use your protractor to draw an angle of 32° at B.

Extend your lines to cross at C
Check by measuring the angle at C
It should equal 180° − 47° − 32° = 101°

b Draw a triangle DEF with DE = 5 cm, DF = 3.2 cm and angle D = 122°.

Step 1 Draw a sketch.

Step 2 Draw your first line a little longer than 5 cm, then mark off D and E.

Step 3 Use your protractor to draw an angle of 122° at D.

Step 4 Measure 3.2 cm along the line with your ruler (or use your compasses set to 3.2 cm, centre D) to mark the point F. Join FE.

c Draw a triangle PQR with PQ = 3.4 cm, PR = 6.8 cm and QR = 4.3 cm.

Step 1 Draw a sketch. It is a good idea to have the longest side as the base of your triangle. That way you won't run out of space.

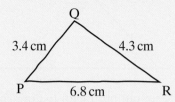

Step 2 Draw your first line a little longer than 6.8 cm, then mark off P and R.

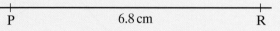

Step 3 Use your compasses with centre P, radius 3.4 cm to draw a big arc.

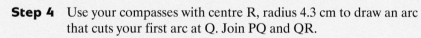

Step 4 Use your compasses with centre R, radius 4.3 cm to draw an arc that cuts your first arc at Q. Join PQ and QR.

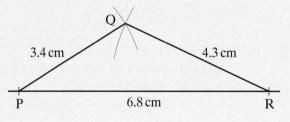

Apply 1

1 Draw these triangles accurately, using the information that you are given.

a

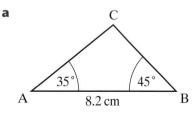

What is the length of AC on your diagram?

b Triangle PQR, with angle P = 72°, PQ = 3.7 cm, PR = 6.3 cm.

Measure and write down angle R.

c Triangle DEF, with DE = 7.1 cm, EF = 4.8 cm, DF = 6.4 cm.

Measure and write down angle F.

2 Make an accurate drawing of a rectangle in which the length of each of the longer sides is 8.4 cm and the length of each diagonal is 9.1 cm. What is the length of each of the shorter sides?

3 Make an accurate drawing of a rhombus in which all the sides are 6 cm and the length of the shorter diagonal is 7 cm. What is the length of the longer diagonal?

4 Get Real!

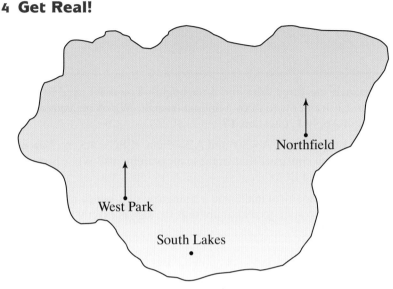

By measuring on the map, work out:

a the bearing of South Lakes from West Park

b the bearing of West Park from Northfield.

The map has a scale of 1 cm to 10 km.

c Find the distance from Northfield to South Lakes to the nearest kilometre.

5 Get Real!
Bhavika is writing a book about making kites.
She makes a sketch of her own kite.

Copy the sketch. Calculate the missing angles and write them on your sketch.

Using a scale of 1 : 15, make an accurate scale drawing of Bhavika's kite.

What is the width of Bhavika's full-size kite?

6 Get Real!
Brian is making a model of his sailing boat. On the actual boat the mainsail is a right-angled triangle. Its vertical height up the mast is 7.20 m. Its hypotenuse is 8.55 m.
Using a scale of 1 : 90, draw accurately a scale plan for the mainsail of the model boat.
By using your scale plan, calculate the width of the mainsail on the full-size boat.

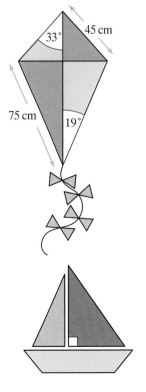

7 Get Real!

On a map the churches of St Clement and St Mary are 9.6 cm apart, with St Clement being due west of St Mary. Use this information to draw the position of the two churches accurately.

The church of St Chad is on a bearing of 102° from St Clement and on a bearing of 248° from St Mary.

Mark the position of St Chad on your map.

If the map scale is 1 : 50 000, what is the distance from St Chad to St Clement in kilometres?

Explore

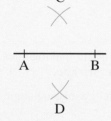

- Every triangle has three sides and three angles. You don't need to know all of these measurements to draw a unique triangle. Which measurements did you need in Apply **1**, question **1**?

- Triangle ABC has angle A = 30° and AB = 5 cm. If BC = 4.5 cm, how many triangles can you construct that fit these measurements? What happens for different lengths of BC?

- If you want a company to make you a triangular sign, which measurements would you need to state to make sure that the sign is exactly the right size and shape?

> Investigate further

Learn 2   Bisecting lines and angles

Examples:

a Construct the bisector of the line AB.

'Construct' means you must produce an accurate diagram using only your compasses and a straight edge (ruler)

Always leave in your construction arcs and remember:
NO ARCS – NO MARKS!

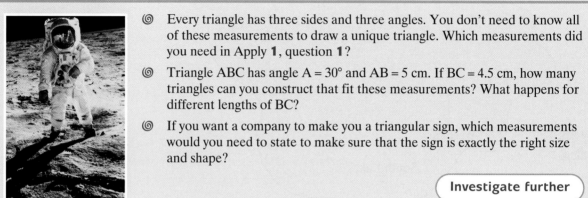

Set the radius of your compasses to more than half of AB.
Put the point on A and draw arcs above and below AB.

Keeping the same radius, put the point of your compasses on B and draw two new arcs to cut the first two at C and D.

Join CD.
X is the midpoint of AB.
CD is not only the bisector of AB, it is the perpendicular bisector.

Use your ruler to check that AX = XB and use your protractor to check that angle CXB = 90° when you have finished your construction

b Construct the bisector of angle BAC.

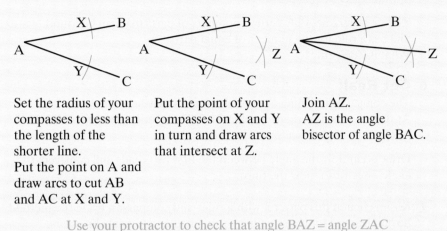

Set the radius of your compasses to less than the length of the shorter line.
Put the point on A and draw arcs to cut AB and AC at X and Y.

Put the point of your compasses on X and Y in turn and draw arcs that intersect at Z.

Join AZ.
AZ is the angle bisector of angle BAC.

Use your protractor to check that angle BAZ = angle ZAC when you have finished your construction

Apply 2

1 **a** Draw the line AB = 9 cm.

b Construct its perpendicular bisector, CD, cutting AB at O.

c Construct the bisector of angle COB.

2 David bisects angle BAC by putting the point of his compasses on B and drawing an arc. Then he puts the point of his compasses on C and, with the same radius as before, he draws a second arc crossing the first. Finally he joins the point where the arcs cut to A.
What has he done wrong?

3 Get Real!
Adam is making a miniature table football game.
The distance between the goal posts is 8 cm.
He needs to mark the penalty spot 12 cm along the perpendicular bisector of the line joining the goal posts.
Construct a diagram showing the positions of the goal posts and penalty spot.

4 Get Real!
Matthew has found a treasure map belonging to the infamous mathematical pirate, Long John Hypotenuse.
It says that the treasure lies somewhere along the bisector of the angle formed by joining the palm tree to Devil's Rock and Devil's Rock to the skull.
Roughly copy the positions of the palm tree, Devil's Rock and the skull. Join them up and construct the angle bisector at Devil's Rock to help Matthew find the path along which the treasure lies.

5 Ben says that he could use his compasses and straight edge to divide a straight line into three equal parts, four equal parts, five equal parts or any number of equal parts.

Zoë says there are only certain numbers of parts that can be found.
She has a rule for the number of equal parts.
What is Zoë's rule?

6 Get Real!

The council is designing a roundabout for a new housing estate.
At the moment two roads meet at 216°. The council plans to add a third road on the angle bisector of the two roads.

Draw accurately the angle between the first two roads and construct the angle bisector to show the third.

7 Kasim created this pattern, with a ruler and compasses, using the method he had learned for constructing perpendicular and angle bisectors.

Can you write out the set of instructions he followed?
Check that your instructions work by seeing what design you create when you follow them.

8 Get Real!

A snooker table measures 357 cm by 178 cm. The black ball lies on a spot 32 cm along the perpendicular bisector of one of the shorter sides.

James is making a scale drawing of a table using a scale of 1 : 20.
Using the same scale, draw one of the shorter sides. Construct its perpendicular bisector and mark the position of the spot for the black ball.

A corner pocket lies at each end of the shorter side. Using your scale drawing, work out the distance between the black ball and one of the corner pockets on the full size snooker table.

Explore

- ◎ Draw a square accurately
- ◎ Construct the midpoints of each side
- ◎ Join them in order
- ◎ What shape have you made?
- ◎ Now repeat the steps starting with a rectangle, trapezium, kite ...

> Investigate further

Explore

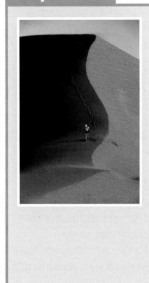

- ◎ Draw a triangle. Construct the perpendicular bisectors of the three sides. They should all meet at one point. This is called the **circumcentre** of the triangle
- ◎ Draw another triangle. Construct the angle bisectors of the three angles. They should all meet at one point. This is called the **incentre** of the triangle
- ◎ Draw another triangle. Construct the midpoints of the three sides. Join each angle to the midpoint of the opposite side. The three lines cross at one point called the **centroid** of the triangle
- ◎ One of the above centres is the centre of gravity of the triangle – that is, the triangle's point of balance. Experiment to find out which centre it is
- ◎ Use the Internet to find out about the other two centres
- ◎ Experiment with different types of triangle
- ◎ Find out how to construct the **orthocentre** of a triangle

> Investigate further

Learn 3 Constructing angles of 90° and 60°

Examples:

a Construct a perpendicular from a point on a line.

Leave in your arcs:
no arcs – no marks!

With the point of your compasses on P, draw two arcs to cut the line either side of P at A and B.	Make the radius of your compasses larger. Put the point on A and B in turn and draw arcs that intersect at C.	Join CP. CP is perpendicular to the original line with a 90° angle at P. Use your protractor to check that angle CPB = 90°.

b Drop a perpendicular from the point P onto the line AB.

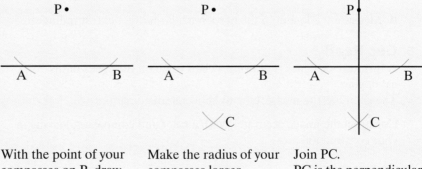

With the point of your compasses on P, draw two arcs that intersect the line at A and B.	Make the radius of your compasses larger. Put the point on A and B in turn and draw arcs that intersect at C.	Join PC. PC is the perpendicular to the original line from the point P.

c Construct an angle of 60°. If you also join RQ you have constructed an equilateral triangle

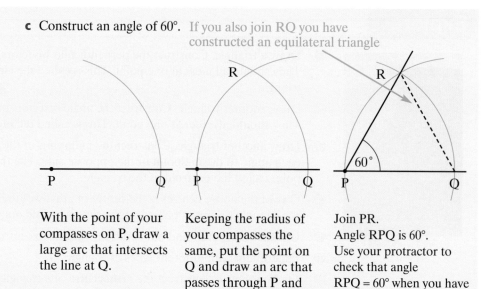

With the point of your compasses on P, draw a large arc that intersects the line at Q.

Keeping the radius of your compasses the same, put the point on Q and draw an arc that passes through P and cuts the first arc at R.

Join PR.
Angle RPQ is 60°.
Use your protractor to check that angle RPQ = 60° when you have finished your construction.

Apply 3

1 Using only a ruler and compasses:

  a construct an angle of 60°

  b bisect your angle to make two angles of 30°.

2 Construct a right-angled triangle with an angle of 30°.

3 a Draw a line AB = 10 cm.

  b Mark the point C on the line AB, with AC = 3.5 cm.

  c Construct a line through C perpendicular to AB.

4 a Draw a line AC = 8 cm.

  b Construct the equilateral triangle ABC.

  c Drop a perpendicular from the point B onto the line AC.

  d Measure the length of the perpendicular to the nearest millimetre.

5 Get Real!
A furniture designer is drawing a chair to go on the cover of his catalogues.
The chair has the measurements shown in the diagram.

Construct the image accurately with a ruler and compasses, leaving in your construction arcs.

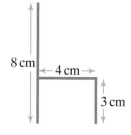

<u>6</u> Get Real!

The diagram shows a roof truss design for a new house.

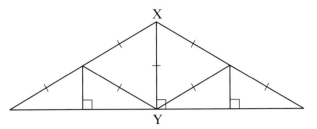

Construct the diagram, starting with the vertical line XY = 3 cm.

Measure the width of the base of the roof truss to the nearest millimetre.

<u>7</u> Get Real!

Draw a circle of radius 5 cm and then draw in a diameter.
Construct a circle divided into 12 equal sectors like the clock face in
the picture.

<u>8</u> Get Real!

Sarah wants to know how far away a pylon is from the hedge at the
boundary of her property.
She marks two points on the hedge, A and B, that are 75 metres apart.
Then she measures the angles at A and B between the hedge and the
lines of sight of the pylon.
The angle at A is 60° and the angle at B is 30°.

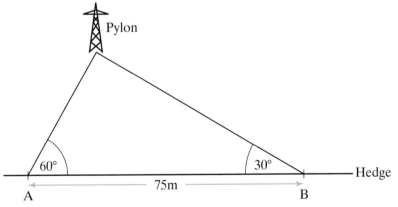

Using a ruler and compasses, construct a scale diagram of Sarah's results,
using a scale of 1 : 1000.
By constructing an extra line, find the shortest distance of the pylon from
Sarah's hedge.

9 Get Real!

Hasan has designed a diamond logo for a bank.

He draws a circle of radius 5 cm, centre O. Then he draws a diameter, AB.

Next he constructs the perpendicular bisector of the diameter and labels it CD. Then he marks a point E on OD, 3 cm from O.
He constructs the perpendicular to OD through E and extends it to F and G on the circumference of the circle.
He repeats the last step for a point H on OC, 3 cm from O, creating the points I and J.
Finally, he joins IF and GJ.
Now he can mark in the diamond shape.

Follow Hasan's design instructions.
What is the width of the diamond shape that you have constructed?

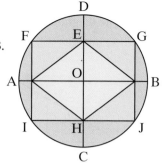

Explore

- ◎ You know how to construct angles of 90° and 60° using only a ruler and compasses
- ◎ Can you construct angles of 30° and 45°?
- ◎ What about 150° and 330°?
- ◎ What other angles can you construct accurately?

Investigate further

Explore

- ◎ A group of students are studying the patterns that are all around us, such as snowflakes, rose windows, clocks, fractals, etc.
- ◎ They start to create some of their own designs

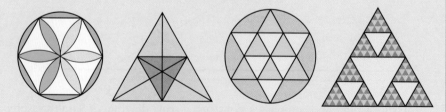

- ◎ Can you use the constructions you have learned to recreate one of the patterns?
- ◎ What designs of your own can you create using only a ruler and compasses?

Investigate further

Learn 4 3-D solids and nets

Example: Draw the nets of the following solids:

a Cuboid

b Triangular prism

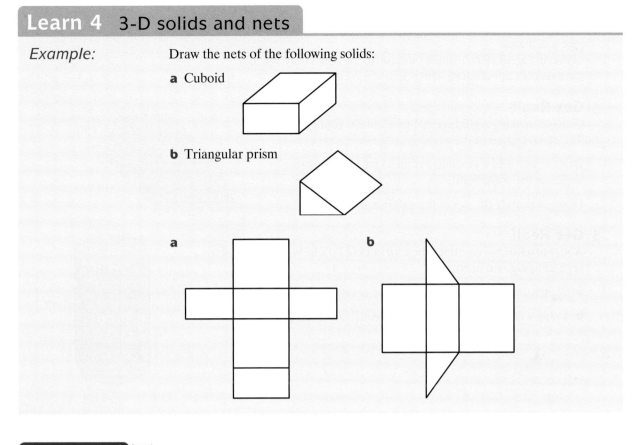

Apply 4

Apart from question **1**, this is a non-calculator exercise.

1 Get Real!

This is a picture of an open box for drawing pins.

a Draw accurately the net of the box.

b Calculate the area of cardboard used to make the box.

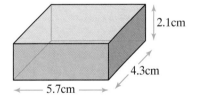

2 This is a diagram of the net for a square-based pyramid.

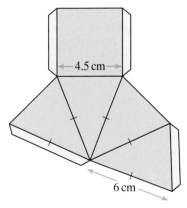

a Construct, using a ruler and compasses only, an accurate full-size net on cardboard.

b Score along the edges and make a model pyramid.

3 Get Real!

William and Robert have invented a game that needs two fair four-sided dice.
Construct accurately, using a ruler and compasses only, the net for a regular tetrahedron with edges of 1.8 cm.

4 Get Real!

Ciarán is making a gift box that looks like a gold bar.
The cross-section of the prism is an isosceles trapezium.

Draw a sketch of a possible net, marking in the missing angles of the trapezium.

Draw accurately the net of this gift box.

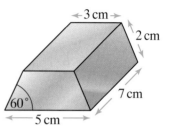

5 Get Real!

A manufacturer is creating a box for cocktail sticks in the shape of a regular hexagonal prism.

a Sketch a net for the box.

b Using a ruler and compasses only, construct a full-size net of the box.

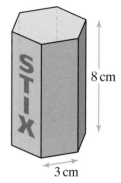

Explore

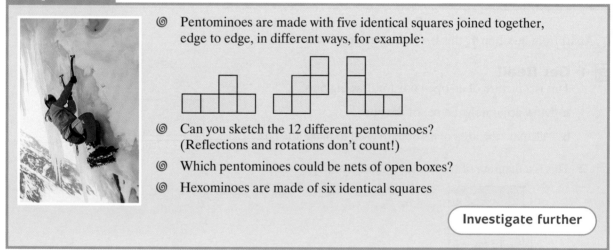

◎ Pentominoes are made with five identical squares joined together, edge to edge, in different ways, for example:

◎ Can you sketch the 12 different pentominoes? (Reflections and rotations don't count!)

◎ Which pentominoes could be nets of open boxes?

◎ Hexominoes are made of six identical squares

Investigate further

Explore

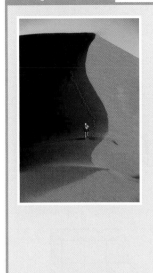

◎ Platonic solids are 3-D solids where all the faces are identical regular polygons

◎ A regular tetrahedron has four equilateral triangles as its faces and a cube has six squares as its faces, so these are both Platonic solids

◎ There are three other Platonic solids – can you find out what they are?

◎ Construct the nets of the solids (you will probably need to use a protractor for one of them)

Investigate further

Learn 5   2-D representations of 3-D solids

Example:

Eight centimetre cubes are glued together to make the 3-D solid below. On a centimetre grid, draw the plan of the solid (viewed from A) and its front and side elevations, viewed from B and C.

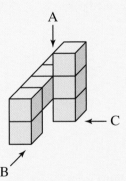

| Plan | Front elevation | Side elevation |
| viewed from A | viewed from B | viewed from C |

Architects and engineers use plans and elevations to represent 3-D solids
A *plan* shows the view of an object from above
An *elevation* shows the object viewed from the front or side

Apply 5

1 Each of the 3-D solids below is made from six centimetre cubes.
On a centimetre grid, draw the plan of each solid and its elevation from
the directions marked F and S.

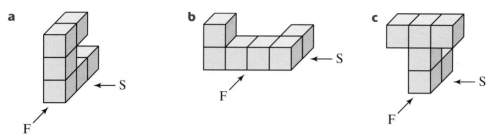

a

b

c

2 Nine centimetre cubes are arranged into a three by three square block,
and another three cubes are placed in front as shown in the diagram.

On a grid draw:

a the plan of this 3-D solid

b the front elevation as viewed from A

c the side elevation as viewed from B.

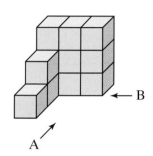

3 Get Real!
The diagram shows the dimensions of a small can of baked beans.
Draw an accurate plan and elevation. Leave out unseen edges.

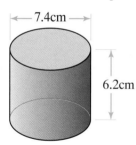

7.4cm

6.2cm

4 For each prism, sketch the plan, front and side elevations from the
directions shown by the arrows. Remember to show unseen edges as
dotted lines.

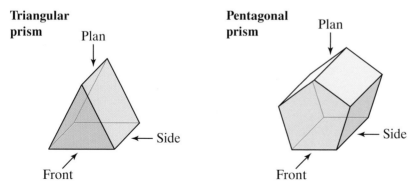

Triangular prism

Plan

Side

Front

Pentagonal prism

Plan

Side

Front

5 Each solid is a cuboid with part removed. The dimensions are given in centimetres.
For each solid, draw a full-size plan and front and side elevations on a centimetre grid.

a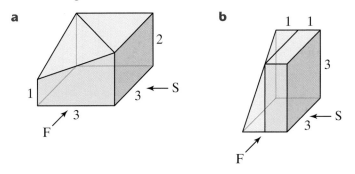

b

6 Sara has drawn a plan and a side elevation (from S) of the 3-D solid shown below.

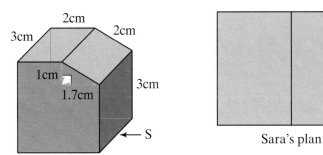

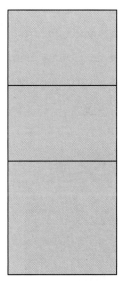

Sara's plan

Sara's side elevation

a Describe what is wrong with Sara's diagrams.

b Draw an accurate plan and side elevation.

7 Plans and elevations of three objects are shown below.
Sketch each object.
You do not need to show hidden edges.

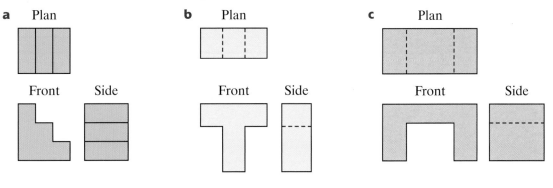

8 Get Real!

The top of this stool is in the shape of a regular hexagon.
It has a cylindrical leg under each corner of the top.

a Sketch a plan of the stool. Show unseen edges with dotted lines.

b Sketch an elevation of the stool when it is viewed from the front, F.

c Sketch an elevation of the stool when it is viewed from the side, S.

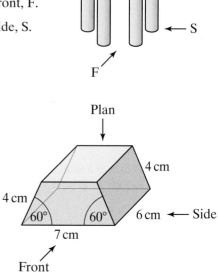

9 The cross-section of a prism is an isosceles trapezium with two equal sides of length 4 cm, a base of length 7 cm and base angles of 60° as shown. The prism is 6 cm long.

Using compasses and ruler only, construct:

a an accurate front elevation of the prism

b an accurate plan of the prism

c an accurate side elevation of the prism.

Explore

◎ Copy and complete the table

Solid	Number of faces	Number of vertices	Number of edges
Cube			
Cuboid			
Square-based pyramid			
Triangular prism			
Pentagonal prism			

◎ Add some other 3-D solids

Investigate further

Construction

ASSESS

The following exercise tests your understanding of this chapter, with the questions appearing in order of increasing difficulty.

1 Two dogs are taking a small boy for a walk!
Scamp's lead is 1.5 m long and Rover's lead is 1 m long.
Unfortunately, they are pulling in different directions at an angle of 40° to each other.
Construct the triangle and measure the distance between Scamp and Rover.

2

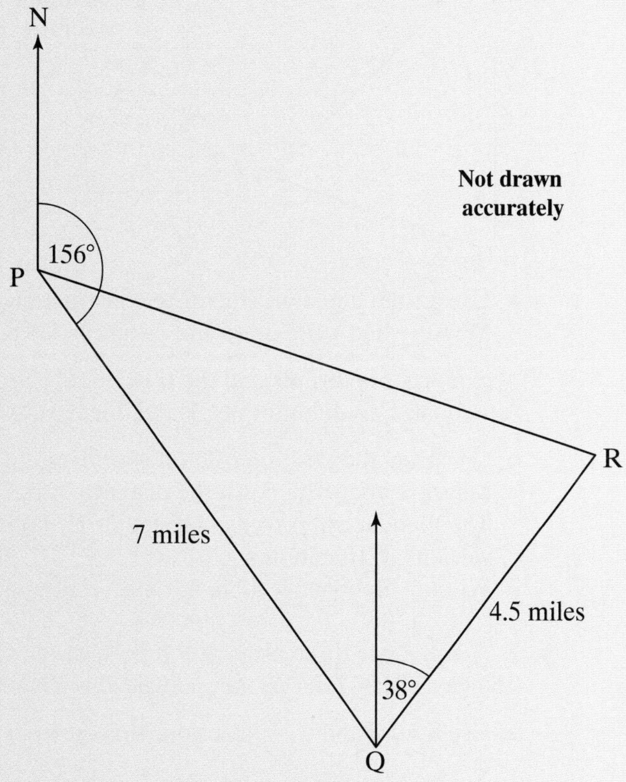

Not drawn accurately

A yacht is taking part in a race.
It leaves the start line at P and has to go round two buoys, Q and R, before sailing back to P.
Q is 7 miles from P on a bearing of 156°.
R is 4.5 miles from Q on a bearing of 038°.

a Draw a scale diagram of the triangle PQR.
Use a scale of 1 cm to 1 mile.

b The yacht has just rounded R.
  i How far does it have to sail back to P?
  ii What is the bearing of R from P?

3 A and B are two lighthouses on a stretch of coastline.
A is 15 miles due north of B.
A trawler, T, is out at sea.
The bearing of T from A is 130°.
The bearing of T from B is 075°.

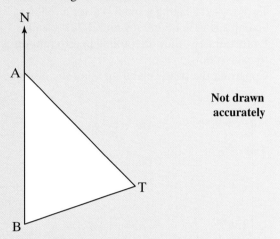

Not drawn
accurately

a Use a ruler and a protractor to construct the triangle ABT.
Measure and write down the distances TA and TB.

A sudden storm erupts and the trawler gets into difficulties.
A lifeboat, L, is due south of T and due east of B.

b Construct the position of L on your diagram.
Measure and write down the distance of L from T.
The lifeboat can travel at a speed of 25 mph and the trawler
will sink in 10 minutes.
Will the lifeboat get there in time?

4 X, Y and Z are three observation posts at the vertices of a right-angled
triangle, where YZ = 20 miles, angle Z is 30° and angle Y is 90°.

a Use a ruler and compasses only to construct this triangle.

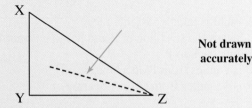

Not drawn
accurately

Observers at X and Z simultaneously record a UFO equidistant from them
over the midpoint of XZ and moving along its perpendicular bisector.
Three seconds later the UFO is lying on the angle bisector of angle Z.

b Using a ruler and compasses only locate the position of the UFO at
this time.

c Measure and write down the distance the UFO has travelled.

d Estimate the speed of the UFO in mph.

5 The cross-section of a prism is a right-angled triangle as shown below.

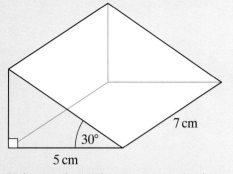

Using a ruler and compasses only, draw an accurate net of the prism.

6 The diagram shows a solid made from eleven one centimetre cubes.

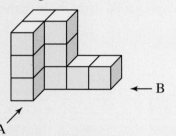

On a centimetre grid draw:

a the plan of this solid

b the front elevation from the direction of the arrow A

c the side elevation from the direction of the arrow B.

Try some real past exam questions to test your knowledge:

7 The kite PQRS is sketched below.
QR = SR = 6 cm
Angle QRS = 50°
The diagonal PR = 8 cm

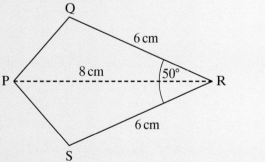

Make an accurate drawing of the kite PQRS.

Spec B, Module 5, Int. Paper 1, Nov 04

8 The diagram shows a prism with an L-shaped cross-section.

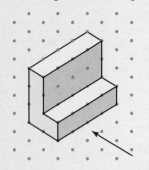

Draw the elevation of this solid, from the direction shown by the arrow.

Spec B, Module 5, Int. Paper 2, Nov 04

9 The diagram shows a solid shape made from 8 cubes.

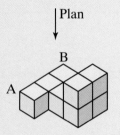

Draw the plan of the solid labelling the points A and B.

Spec A, Int. Paper 1, Nov 03

8 Probability

What you should already know ...

 Use a probability scale

 The difference between experimental and theoretical probability

 Calculate simple probabilities

 Find outcomes systematically

Probability – a value between 0 and 1 (which can be expressed as a fraction, decimal or percentage) that gives the likelihood of an event

Outcome – the result of an experiment, for example, when you toss a coin, the outcome is a head or a tail

Certain – an outcome with probability 1, for example, the sun rising and setting

Likely – an outcome with a probability greater than $\frac{1}{2}$, for example, rain falling in November in the UK

Unlikely – an outcome with a probability less than $\frac{1}{2}$, for example, snow falling in August in the UK

Impossible – an outcome with a probability 0, for example, the sun turning green

Evens – probability $\frac{1}{2}$, for example, there is an even chance of getting a head or a tail when you toss a coin

Theoretical probability – probability based on equally likely outcomes, for example, it suggests you will get 5 heads and 5 tails if you toss a coin 10 times

Experimental probability or **relative frequency** – this is found by experiment, for example, if you get 6 heads and 4 tails, the experimental probability would be $\frac{6}{10}$ or 0.6 for getting a head

Random – a choice made when all outcomes are equally likely, for example, picking a raffle ticket from a box with your eyes shut

Mutually exclusive events – these cannot both happen in the same experiment, for example, getting a head and a tail on one toss of a coin

Two-way table or **sample space diagram** – table used to show all the possible outcomes of an experiment, for example, all the outcomes of tossing a coin and throwing a dice

		Dice					
		1	**2**	**3**	**4**	**5**	**6**
Coin	**Head**	H1	H2	H3	H4	H5	H6
	Tail	T1	T2	T3	T4	T5	T6

Independent events – events are independent when the outcome of one does not affect the outcome of the other, for example, tossing a coin and drawing a card from a pack

Dependent events – events are dependent when the outcome of one affects the outcome of the other, for example, taking two successive balls from a box without replacing the first one

Learn 1   Experimental and theoretical probability

Examples:

The experimental and theoretical results of throwing a dice 600 times are shown in the table below.

	Outcome					
	1	**2**	**3**	**4**	**5**	**6**
Theoretical results	100	100	100	100	100	100
Experimental results	92	107	103	99	97	102

The probability of getting a six from one throw of a dice is $\frac{1}{6}$

This is a theoretical probability which leads to the assumption that there will be 100 sixes in 600 throws of the dice

In practice, it is unlikely that there will be exactly 100 of each number

a Use the table to calculate:

i the theoretical probability of getting a 6.

The theoretical probability is $\frac{100}{600} = \frac{1}{6}$

The theoretical probability cancels down to $\frac{1}{6}$

ii the experimental probability of getting a 6.

The experimental probability is $\frac{102}{600}$

The experimental probability is also called the relative frequency

b Do you think the dice is biased? Give a reason for your answer.

An unbiased dice is usually called a fair dice

The dice is not biased as the theoretical and experimental frequencies are close to each other.

Some events cannot be predicted theoretically, for example, the probability that there will be 12 sunny days in August; the data for weather in August over the past five years could be used to find an experimental probability for this outcome

Apply 1

Apart from question 8, this is a non-calculator exercise.

1 A coin is tossed 300 times.

How many times would you expect to get a head?

2 The probability of getting a faulty light bulb is 0.01

How many faulty light bulbs would you expect to find in a batch of 500?

3 A spinner with five equal divisions labelled A, B, C, D, E is spun 100 times.

a How many times would you expect it to land on A?

b How many times would you expect it to land on D?

c Julie spins the spinner 10 times and gets 2 As, 3 Bs, no Cs and 5 Ds.

She says this shows the spinner is biased. Is she correct?

Give a reason for your answer.

4 The probability that an inhabitant of Random Island has red hair is $\frac{3}{20}$

There are 4263 people living on Random Island.

How many of them would you expect to have red hair?

5 The table shows the frequency distribution after taking a letter at random from the word ASSESS 60 times.

	Results from drawing a letter 60 times		
	A	**E**	**S**
Frequency	13	10	37

a What is the relative frequency of getting the letter S?

b How does this compare with the theoretical probability of getting the letter S?

c If this experiment is repeated 600 times, how many times would you expect to get the letter E?

6 Kali has a spinner with coloured sections of equal size.
She wants to know the probability that her spinner lands on pink.
She spins it 100 times and calculates the relative frequency of pink after
every 10 spins.
Her results are shown on the graph.

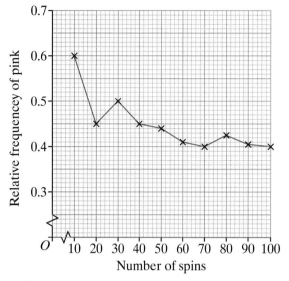

a Use the graph to calculate the number of times that the spinner
landed on pink:

  i after the first 10 spins

  ii after the first 50 spins.

b From the graph, estimate the probability of the spinner landing on pink.

c Kali's results confirm that her spinner is fair.
The spinner has five equal sections.

  i How many sections are pink?

  ii Kali spins the spinner two more times.
  What is the theoretical probability that the spinner lands on pink
  both times?

7 A spinner is made from a regular octagon.
Each of the eight sections of the spinner has one of the letters P, Q, R, S.
It is spun 240 times and the results are shown in the table.

Results from 240 spins			
P	Q	R	S
Frequency	34	62	88	56

How many times does each letter appear on the spinner?

8 Emma has three dice – one is red, one is blue and one is green.
The table shows the frequency distributions after throwing each dice
600 times.

Results from 600 throws of dice						
	1	**2**	**3**	**4**	**5**	**6**
Red dice	111	108	89	98	96	98
Blue dice	54	65	83	121	129	148
Green dice	88	99	106	96	102	109

   a One of the dice appears to be biased. Which one is this?
   What suggests that it is biased?

   b From the two fair dice, what is the relative frequency of throwing a 5?
   How does this compare with the theoretical probability?

   c From the two fair dice, what is the relative frequency of throwing a
   number greater than 3?
   How does this compare with the theoretical probability?

9 Sam has some counters of assorted colours in a bag.
He asks his friends to pick a counter with their eyes shut.
He wants to find the probability of getting a red counter.
After every 10 attempts, he finds the relative frequency of red and the
results are shown on the graph.

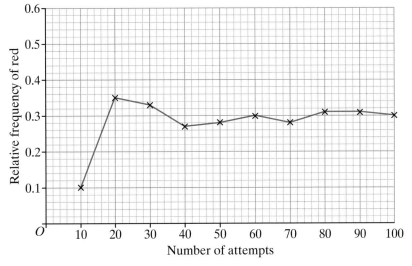

   a How many red counters had been picked after 10 attempts?

   b How many red counters had been picked after 50 attempts?

   c From the graph, estimate the probability of picking a red counter.

   d If there are 12 red counters, how many counters are in the bag altogether?

Explore

- ◉ Throw two dice together and write down the sum of their scores
- ◉ Repeat this until you have recorded 30 totals
- ◉ Which total appears the most?
- ◉ Which total appears the least?

Investigate further

Learn 2 Combined events

Example:

Mollie tosses a coin and picks a letter at random from the word SPACE.
Draw a table to show this information.
What is the probability that she gets a head and a vowel?

	S	P	A	C	E
Head	Head, S	Head, P	Head, A	Head, C	Head, E
Tail	Tail, S	Tail, P	Tail, A	Tail, C	Tail, E

There are 10 different possible outcomes.

There are 2 outcomes with a head and a vowel.

P(head and vowel) = $\frac{2}{10} = \frac{1}{5}$

Remember to cancel down where possible

These results have been set out systematically in a two-way table, so that none are left out

Apply 2

1 The Aces, Kings, Queens and Jacks are taken from a full pack of cards to form a new, smaller pack. A card is then drawn at random from this pack.

		A	K	Q	J
Clubs	♣	A♣	K♣		
Hearts	♥	A♥			
Diamonds	♦	A♦			
Spades	♠				

- **a** Copy and complete the table of outcomes (two-way table) and use it to find the probability of drawing:
- **b** the King of diamonds
- **c** a black Queen
- **d** a red Jack *or* a red Ace
- **e** a black card
- **f** the King *or* Queen of clubs.

2 Alan has five tiles and three cards as shown in the diagram.

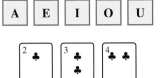

He picks up one tile and one card at random.

a Draw a table of outcomes.

Find the probability that Alan picks:

b tile A and the 3 of clubs

c tile A and a card that is *not* the 3 of clubs

d a tile that is *not* A and a card with an even number.

3 Two dice are thrown and their scores are added together.

a Copy and complete the two-way table.

	1	2	3	4	5	6
1	2		4			
2		4				
3	4					
4						
5						
6						

Use your two-way table to find:

b the probability of getting a total of 4

c the probability of getting a total of 14

d the probability of getting a total of 7

e the probability of getting a total which is at least 5.

4 A box contains 4 green counters and 1 red counter.
Mo chooses one counter at random and Sean throws a dice.

a Draw a two-way table.

b Find the probability that Mo gets a red counter and Sean throws a 5.

c Find the probability that Mo gets a green counter and Sean throws a 6.

d Find the probability that Mo gets a red counter and Sean throws an even number.

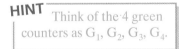

HINT Think of the 4 green counters as G_1, G_2, G_3, G_4.

5 Tim picks a card at random from a full pack.
Jo picks a marble at random from a bag containing 2 black marbles and 3 red marbles. Find the probability that:

a Tim picks a diamond and Jo picks a black marble

b Tim's card is from a red suit and Jo's marble is also red

c Tim does not pick a diamond and Jo picks a red marble.

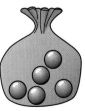

6 Two dice are thrown and the product of their scores is recorded.

   a Copy and complete the two-way table.

	1	2	3	4	5	6
1	1					
2			6			
3						
4						
5					25	
6						

Use your two-way table to find:

   b the probability of getting a product of 15

   c the probability of getting a product of 12

   d the probability of getting a product greater than 16

   e the probability of getting a product that is an even number.

7 Lucy says that if you toss two coins, the possible outcomes are two heads, two tails or one of each. She therefore decides that the probability of getting two heads is $\frac{1}{3}$

Explain why Lucy is wrong.

8 Shaun tosses three coins.

   a How many possible outcomes are there?

   b What is the probability that he gets one head and two tails?

   c What is the probability that he gets at least one head?

Explore

  ◎ How many possible outcomes are there from tossing 2 coins?

  ◎ How many possible outcomes are there from tossing 3 coins?

  ◎ How many possible outcomes are there from tossing 4 coins?

  ◎ How many possible outcomes are there from tossing 5 coins?

                                        Investigate further

Learn 3   Mutually exclusive events

Examples:

a Are the outcomes 'throwing a 6' and 'throwing a 2' mutually exclusive?

If you throw a dice once, you may get a 6 and you may get a 2.

You cannot get both a 6 and a 2 on the same throw.

The outcome 'throwing a 6' and the outcome 'throwing a 2' are mutually exclusive.

b Are the outcomes 'drawing a diamond' and 'drawing a 6' mutually exclusive?

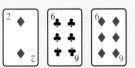

If you draw one card from a pack, you may get a diamond and you may get a 6.

You could get the 6 of diamonds, so the outcome 'drawing a diamond' and the outcome 'drawing a 6' are *not* mutually exclusive.

> The sum of the probabilities of all the mutually exclusive events is 1

If event A and event B are mutually exclusive, then P(A or B) = P(A) + P(B).

c There are 7 white counters, 8 black counters and 5 yellow counters in a bag. One is drawn at random.
Find the probability of getting:

i a white counter

ii a black counter

iii a yellow counter

iv a black or a yellow counter

v a counter that is not white.

Drawing a white counter, drawing a black counter and drawing a yellow counter are mutually exclusive events.

> Remember to cancel down where possible

i $P(\text{white}) = \frac{7}{20}$

ii $P(\text{black}) = \frac{8}{20} = \frac{2}{5}$

iii $P(\text{yellow}) = \frac{5}{20} = \frac{1}{4}$

> Adding these probabilities we get $\frac{7}{20} + \frac{8}{20} + \frac{5}{20} = \frac{20}{20} = 1$
> Use this as a check on your work

iv $P(\text{black or yellow}) = P(\text{black}) + P(\text{yellow}) = \frac{8}{20} + \frac{5}{20} = \frac{13}{20}$

v $P(\text{not white}) = 1 - \frac{7}{20} = \frac{13}{20}$

> This is the same as P(black or yellow)

Apply 3

1 Which of these are mutually exclusive events?

   a Getting a six and an odd number on a throw of a dice.

   b Getting a four and an even number on a throw of a dice.

   c Picking a heart and the Ace of clubs from a pack of cards.

   d Picking the Queen of diamonds and the Jack of diamonds from a pack of cards.

   e Picking a club and a King from a pack of cards.

2 The probability that Nicola will come top in the maths exam is 0.8
What is the probability that she will not come top?

3 The probability that Andy will fail his driving test is $\frac{4}{15}$
What is the probability that he will pass?

4 The probability that Random United will win their next match is 0.3
The probability that they will draw is 0.6
What is the probability that they will lose?

5 The probability that Steve will score more than 50% in his history test is 0.7
Jenny says, 'That means there is a probability of 0.3 that he will get less than 50%'.
Explain why she is wrong.

6 The table shows the probabilities that a student at Random High School will choose certain lunch menus.

Salad	Pizza	Hotpot	Other
0.2	...	0.1	0.3

What is the probability that the student will choose pizza?

7 There are three possible transport routes from Canary Wharf to Waterloo: a waterbus, the Docklands Light Railway (DLR) or the Jubilee Line.
The probability that Andy uses the waterbus is 0.16
The probability that he uses the DLR is 0.4
Helen says this means there is a probability of 0.8 that he uses the Jubilee Line.
What mistake has she made? What is the correct probability?

8 The table shows the probabilities that the next car going past the school will be a particular colour.

Silver	Red	Black	White	Other
0.34	0.15	0.14	...	0.3

   a What is the probability that the next car will be white?

   b What is the probability that the next car will be either red or black?

9 Errol goes out to buy a new sweatshirt.
The probability that he buys it from Supershirts is $\frac{9}{20}$
The probability that he buys it from BestSweats is $\frac{2}{5}$
What is the probability that he buys it from one of these stores?

10 The probability that Zoe will move to Rugby is 0.4
The probability that she will move to Coventry is 0.2
The probability that she will move to Leicester is 0.3

  a What is the probability that Zoe will move to either Rugby or Leicester?

  b What is the probability that she will *not* move to Coventry?

  c What is the probability that she will not move to any of these three towns?

11 Jayne has a collection of 50 old teapots.
27 of them have rose patterns. Some of them are covered in blue willow pattern. Some of them have striped patterns. The rest have plain colours. One teapot is chosen at random.

The probability that it will have a striped pattern is 0.08

  a How many of Jayne's teapots have a striped pattern?

The probability that it will have a blue willow pattern is 0.22

  b How many of Jayne's teapots have plain colours?

12 Matt has 25 books on his bookshelf. Some of them are history textbooks, some are science fiction, five are biographies and two are dictionaries. He has put his library card in one of the books but cannot remember which one.

  a Find the probability that the library card is in either a biography or a dictionary.

The probability that it is in a history textbook is 0.4.

  b How many science fiction books are on the shelf?

Learn 4   Independent events

Example: Kathy tosses a coin and Liam draws a card from a pack at random.
What is the probability that Kathy throws a head and Liam draws an Ace?

Draw a two-way table:

Liam

		A	2	3	4	5	6	7	8	9	10	J	Q	K
Kathy	**H**	HA	H2	H3	H4	H5	H6	H7	H8	H9	H10	HJ	HQ	HK
	T	TA	T2	T3	T4	T5	T6	T7	T8	T9	T10	TJ	TQ	TK

The probability that Kathy tosses a head and Liam draws an Ace is $\frac{1}{26}$

Alternatively P(head) = $\frac{1}{2}$

        P(ace) = $\frac{1}{13}$

P(head and ace) = $\frac{1}{2} \times \frac{1}{13} = \frac{1}{26}$

If the events are independent then the probability that they will both happen is found by multiplying the probabilities together

If events A and B are independent then P(A and B) = P(A) × P(B)

Apply 4

1 Anna throws a dice and Bob draws a card at random.
What is the probability that:

   a Anna gets a 6 and Bob gets a King

   b Anna gets an odd number and Bob gets a diamond

   c Anna gets a number greater than 2 and Bob gets a red 5?

2 The probability that Jo will come top in maths is 0.3
The probability that she will come top in French is 0.4
What is the probability that she comes top in both subjects?

3 The probability that Steve has porridge for breakfast is $\frac{1}{4}$
The probability that he catches a bus to work is $\frac{3}{5}$
What is the probability that Steve has porridge and catches the bus to work?

4 The probability that Paul washes his car on Saturday is $\frac{1}{6}$
The probability that he rents a video is $\frac{2}{9}$
What is the probability that Paul does not wash his car but does rent a video?

5 Lorraine has a packet of crisps and a fruit juice for lunch every day.
The probability that she has roast chicken crisps is 0.2
The probability that she has orange juice is 0.4
What is the probability that she has roast chicken crisps and orange juice?

6 Two dice are thrown together.
What is the probability that the combined score is 2?

7 Three dice are thrown together.
What is the probability that the combined score is 18?

8 The probability that Lucy will go to Spain for her holiday is 0.55
The probability that she will go on holiday in July is 0.44
Marie says that means Lucy is 99% certain to go to Spain in July.
Explain why Marie is wrong.

9 Dee, Karen and Matt are all taking their driving tests next week.
The probability that Dee will pass is 0.7
The probability that Karen will pass is 0.9
The probability that Matt will pass is 0.6
The results of their tests are independent of each other.

   a What is the probability that all three of them will pass?

   b What is the probability that Dee and Karen will pass and Matt will fail?

   c What is the probability that all three of them will fail?

10 The probability that Random United will win their next match is $\frac{7}{12}$
The probability that AQA Rovers will win their next match is $\frac{3}{7}$
They are not playing each other in their next match.

   a What is the probability that both teams will win?

   b What is the probability that neither team will win their match?

Learn 5   Tree diagrams

Example:

A bag contains 4 black counters and 5 white counters.
Two counters are taken at random from the bag.
Draw a tree diagram to show the information.

'Replaced' tells you that the second event is independent of the first event

The answer depends on whether the first counter is replaced (independent events) or not replaced (dependent events).

The possibilities are shown separately below.

i With replacement (independent events)
One counter is taken at random from the bag and then replaced.
A second counter is then taken at random from the bag.

The two events are *independent* because the colour of the first counter does not affect the outcome when the second counter is taken

A tree diagram is useful for calculating probabilities
The probabilities are written on the branches of the tree

The tree diagram is shown below.

1st counter   2nd counter

$\frac{4}{9}$ B — B $P(B, B) = \frac{4}{9} \times \frac{4}{9} = \frac{16}{81}$

$\frac{5}{9}$ — W $P(B, W) = \frac{4}{9} \times \frac{5}{9} = \frac{20}{81}$

$\frac{5}{9}$ W — $\frac{4}{9}$ B $P(W, B) = \frac{5}{9} \times \frac{4}{9} = \frac{20}{81}$

$\frac{5}{9}$ — W $P(W, W) = \frac{5}{9} \times \frac{5}{9} = \frac{25}{81}$

The first counter was replaced so it did not affect the outcome (or the probabilities) when the second counter was taken out

The two branches always add up to one, because they show all the possible outcomes

ii Without replacement (dependent events)
One counter is taken at random from the bag and not replaced.
A second counter is then taken at random from the bag.

The two events are *dependent* because the colour of the first counter does affect the outcome when the second counter is taken

'Not replaced' tells you the second event is dependent on the first event

1st counter   2nd counter

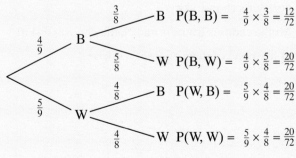

$\frac{4}{9}$ B — $\frac{3}{8}$ B $P(B, B) = \frac{4}{9} \times \frac{3}{8} = \frac{12}{72}$

$\frac{5}{8}$ — W $P(B, W) = \frac{4}{9} \times \frac{5}{8} = \frac{20}{72}$

$\frac{5}{9}$ W — $\frac{4}{8}$ B $P(W, B) = \frac{5}{9} \times \frac{4}{8} = \frac{20}{72}$

$\frac{4}{8}$ — W $P(W, W) = \frac{5}{9} \times \frac{4}{8} = \frac{20}{72}$

The first counter was not replaced so it affected the outcome when the second counter was taken out

If the first counter is black, there are only 3 black counters left and only 8 counters in the bag. So the probability of getting a second black counter becomes $\frac{3}{8}$

There are still 5 white counters, so the probability of getting a white counter becomes $\frac{5}{8}$

Apply 5

1 A box contains 3 red pencils and 7 green pencils. A pencil is taken from the box and then replaced. A second pencil is then taken from the box.

a Copy and complete the tree diagram.

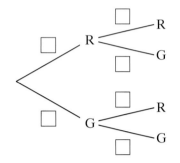

Use the tree diagram to find:

b the probability that both pencils are red

c the probability that both pencils are green

d the probability that one pencil is red and one is green.

2 Sam has five sweaters – two grey and three tan. He has four pairs of jeans – three blue pairs and one black pair. On Saturday morning he chooses a sweater and a pair of jeans at random.

a Draw a tree diagram.

Use the tree diagram to find:

b the probability that Sam chooses a grey sweater and black jeans

c the probability that he chooses a tan sweater and blue jeans

d the probability that he chooses a grey sweater and blue jeans.

3 Girls at Random High School can choose to wear a navy skirt or navy trousers.
The probability that Mollie will choose a skirt is 0.4 and the probability that her friend Nadia will choose a skirt is 0.7

a Draw a tree diagram.

Use the tree diagram to find:

b the probability that both girls choose skirts

c the probability that Mollie chooses trousers and Nadia chooses a skirt

d the probability that at least one of them chooses a skirt.

4 The probability that Rajesh will oversleep is 0.2
If he oversleeps, the probability that he will be late for school is 0.8
If he does not oversleep, the probability that he will be late is 0.1

  a Draw a tree diagram.

  Use the tree diagram to find:

  b the probability that Rajesh does not oversleep and is on time for school

  c the probability that he is late for school.

5 The probability that George will play cricket after school is $\frac{2}{5}$
If he plays cricket, the probability that he will forget his homework is $\frac{5}{8}$
If he does not play cricket, the probability that he will forget his homework is $\frac{1}{4}$

  a Draw a tree diagram.

  Use the tree diagram to find:

  b the probability that George plays cricket and forgets his homework

  c the probability that he does not play cricket and remembers his homework

  d the probability that he forgets his homework.

6 Amy has five pound coins and seven 20p pieces in her purse.
She takes out two coins at random. (Think of this as taking first one coin, then another, without replacement.)

  a Draw a tree diagram.

  Use the tree diagram to find:

  b the probability that Amy takes two 20p pieces

  c the probability that she takes two pound coins

  d the probability that she takes at least one pound coin.

7 Two cards are drawn at random from a full pack.

  a Draw a tree diagram.

  Use the tree diagram to find:

  b the probability that both cards are Aces

  c the probability that neither card is an Ace

  d the probability that just one card is an Ace.

8 Dean has 8 black socks, 6 grey socks and 4 red socks in his drawer.
He takes out one sock and then another.

a Copy and complete the tree diagram.

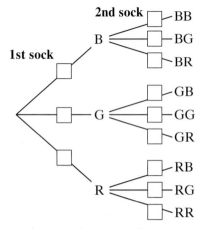

Use the tree diagram to find:

b the probability that Dean has picked two black socks

c the probability that Dean has picked one grey and one red sock

d the probability that Dean has picked a pair of matching socks.

9 The probability that Kylie will pass her driving test is $\frac{1}{3}$
This probability does not change, however many times she takes the test.
Find the probability that:

a Kylie passes her test at the second attempt

b Kylie passes her test at the fifth attempt

c Kylie passes her test at the nth attempt.

Probability

The following exercise tests your understanding of this chapter,
with the questions appearing in order of increasing difficulty.

1 Two pentagonal spinners, each with the numbers 1 to 5, are spun and their
outcomes added together to give a score.

a Draw a two-way table for the two spinners.

b Use your table to find:

i the probability of a score of 4

ii the probability of a score of 5

iii the probability of a score of 9

iv the most likely score.

2 a Which of these pairs of events are mutually exclusive events?

    i Throwing a 3 and an even number on a throw of a dice.

    ii Throwing a 1 and an odd number on a throw of a dice.

    iii Picking a spade and a club from a pack of cards.

    iv Picking a diamond and a King from a pack of cards.

b The probability that a train arrives early is 0.09
The probability that it arrives late is 0.4
What is the probability that it arrives on time?

c The table shows the probabilities of selecting tickets from a bag. The tickets are coloured yellow, black or green and numbered 1, 2, 3 or 4.

	1	2	3	4
Yellow	$\frac{1}{20}$	$\frac{1}{16}$	$\frac{3}{40}$	$\frac{1}{8}$
Black	$\frac{1}{10}$	$\frac{3}{40}$	0	$\frac{3}{40}$
Green	0	$\frac{1}{8}$	$\frac{3}{16}$	$\frac{1}{8}$

A ticket is taken at random from the bag. Calculate the probability that:

    i it is black and numbered 4

    ii it is numbered 2

    iii it is green

    iv it is yellow or numbered 1.

3 Nazeem selected one pen from a box containing 3 red, 4 green and 1 blue pens and a second pen from another box containing 2 red and 1 green. Copy and complete the tree diagram to show the possible outcomes.

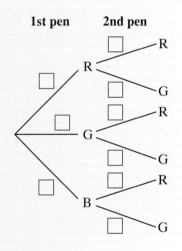

4 a Use your tree diagram from question **3** to find the probability of each possible outcome.

b Which outcome is the:

    i least likely to occur   **ii** most likely to occur?

c Nazeem actually selected two pens of the same colour. Which colour is she most likely to have selected?

d Find the probability that the two pens selected were:

    i the same colour   **ii** different colours.

5 In Class 11A at Thornes Comprehensive there are 21 girls and 9 boys. Everyone in the school has hair that can be classed as blonde, brown or black and the probability of having hair of that colouring is:

Blonde	0.2
Brown	0.3
Black	0.5

a Draw a tree diagram which shows all six possible outcomes for gender and hair colouring.

b One student is chosen at random from the class. Calculate the probability that this student is:

i a boy with brown hair

ii a girl with black hair

iii not blonde.

9 Vectors

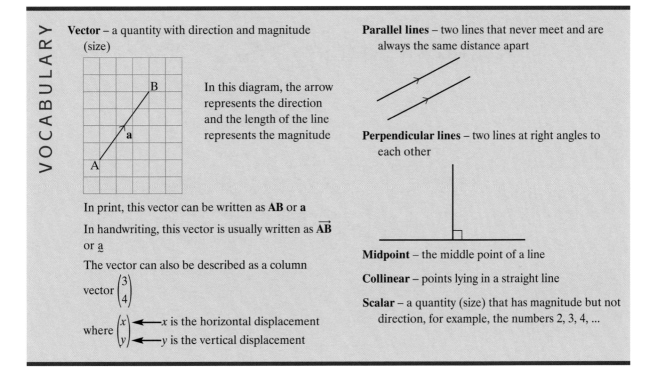

OBJECTIVES

A ▶ **Examiners would normally expect students who get an A grade to be able to:**

Add, subtract and multiply vectors to solve vector geometry problems

Understand the relationship between parallel and perpendicular vectors

A* ▶ **Examiners would normally expect students who get an A* grade also to be able to:**

Solve more difficult vector geometry problems

What you should already know ...

■ Write a column vector

■ Translate a shape using a column vector

■ Understand parallel and perpendicular

VOCABULARY

Vector – a quantity with direction and magnitude (size)

In this diagram, the arrow represents the direction and the length of the line represents the magnitude

In print, this vector can be written as **AB** or **a**

In handwriting, this vector is usually written as $\overrightarrow{AB}$ or $\underset{\sim}{a}$

The vector can also be described as a column vector $\begin{pmatrix} 3 \\ 4 \end{pmatrix}$

where $\begin{pmatrix} x \\ y \end{pmatrix}$ ◀— x is the horizontal displacement
◀— y is the vertical displacement

Parallel lines – two lines that never meet and are always the same distance apart

Perpendicular lines – two lines at right angles to each other

Midpoint – the middle point of a line

Collinear – points lying in a straight line

Scalar – a quantity (size) that has magnitude but not direction, for example, the numbers 2, 3, 4, ...

Learn 1 Column vectors: addition, subtraction and multiplication

Examples:

a Make use of the diagram to write the following as column vectors:

  i **a**
  ii **b**
  iii **a + b**
  iv **a − b**

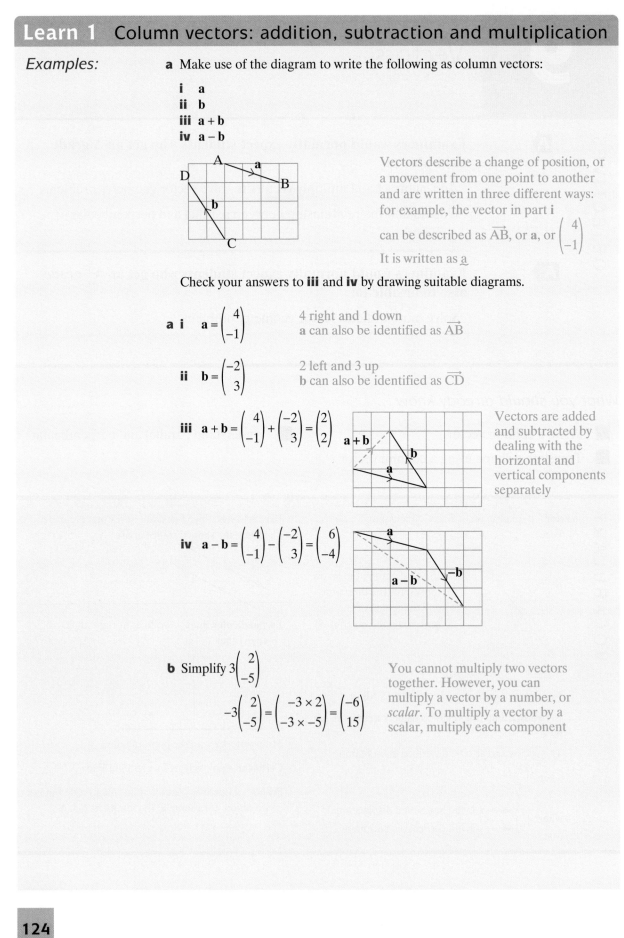

Vectors describe a change of position, or a movement from one point to another and are written in three different ways: for example, the vector in part **i** can be described as $\overrightarrow{AB}$, or **a**, or $\begin{pmatrix} 4 \\ -1 \end{pmatrix}$

It is written as <u>a</u>

Check your answers to **iii** and **iv** by drawing suitable diagrams.

a i $\quad \mathbf{a} = \begin{pmatrix} 4 \\ -1 \end{pmatrix}$ 	4 right and 1 down
a can also be identified as $\overrightarrow{AB}$

ii $\quad \mathbf{b} = \begin{pmatrix} -2 \\ 3 \end{pmatrix}$ 	2 left and 3 up
b can also be identified as $\overrightarrow{CD}$

iii $\quad \mathbf{a} + \mathbf{b} = \begin{pmatrix} 4 \\ -1 \end{pmatrix} + \begin{pmatrix} -2 \\ 3 \end{pmatrix} = \begin{pmatrix} 2 \\ 2 \end{pmatrix}$

Vectors are added and subtracted by dealing with the horizontal and vertical components separately

iv $\quad \mathbf{a} - \mathbf{b} = \begin{pmatrix} 4 \\ -1 \end{pmatrix} - \begin{pmatrix} -2 \\ 3 \end{pmatrix} = \begin{pmatrix} 6 \\ -4 \end{pmatrix}$

b Simplify $3\begin{pmatrix} 2 \\ -5 \end{pmatrix}$

$$-3\begin{pmatrix} 2 \\ -5 \end{pmatrix} = \begin{pmatrix} -3 \times 2 \\ -3 \times -5 \end{pmatrix} = \begin{pmatrix} -6 \\ 15 \end{pmatrix}$$

You cannot multiply two vectors together. However, you can multiply a vector by a number, or *scalar*. To multiply a vector by a scalar, multiply each component

Apply 1

1 Write the vectors shown in the diagram as column vectors.

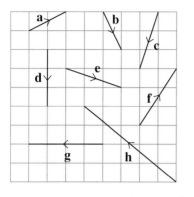

2 a The vectors $\mathbf{m} = \begin{pmatrix} 3 \\ 2 \end{pmatrix}$, $\mathbf{n} = \begin{pmatrix} 4 \\ -2 \end{pmatrix}$, $\mathbf{p} = \begin{pmatrix} -4 \\ 3 \end{pmatrix}$ and $\mathbf{q} = \begin{pmatrix} -3 \\ -3 \end{pmatrix}$.

Calculate:

i $\mathbf{m} + \mathbf{n}$	**iv** $\mathbf{n} - \mathbf{p}$	**vii** $\mathbf{m} + \mathbf{p} + \mathbf{q}$
ii $\mathbf{m} - \mathbf{p}$	**v** $\mathbf{p} - \mathbf{q}$	**viii** $\mathbf{n} - \mathbf{p} - \mathbf{q}$
iii $\mathbf{m} + \mathbf{q}$	**vi** $\mathbf{q} - \mathbf{n}$	**ix** $\mathbf{m} + \mathbf{n} + \mathbf{p} + \mathbf{q}$

b Check your answers with diagrams.

3 Get Real!

In a game of beach volleyball, player A hits the ball with vector $\begin{pmatrix} 1 \\ -5 \end{pmatrix}$.

However, the wind is blowing to add an extra $\begin{pmatrix} 1 \\ 2 \end{pmatrix}$ to the ball's direction.

Does the ball end up in the court or out?

Explain your answer with a diagram.

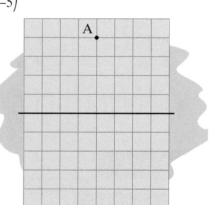

4 If $\mathbf{r}$ is the vector $\begin{pmatrix} -4 \\ 3 \end{pmatrix}$ and $\mathbf{r} + \mathbf{s} = \begin{pmatrix} 2 \\ -1 \end{pmatrix}$, write $\mathbf{s}$ as a column vector.

5 Pierre says that $\begin{pmatrix} 3 \\ -2 \end{pmatrix} - \begin{pmatrix} -2 \\ -1 \end{pmatrix} = \begin{pmatrix} 1 \\ -3 \end{pmatrix}$.

Quinton says that $\begin{pmatrix} 3 \\ -2 \end{pmatrix} - \begin{pmatrix} -2 \\ -1 \end{pmatrix} = \begin{pmatrix} 5 \\ -3 \end{pmatrix}$.

Ravinda says that $\begin{pmatrix} 3 \\ -2 \end{pmatrix} - \begin{pmatrix} -2 \\ -1 \end{pmatrix} = \begin{pmatrix} 1 \\ -1 \end{pmatrix}$.

Who is correct?
Give reasons for your answers.

6 Get Real!

A man rows a boat from the bank at A to point B in the river.
He stops for a rest, and the boat drifts to C. He then rows to D,
before heading back to A.

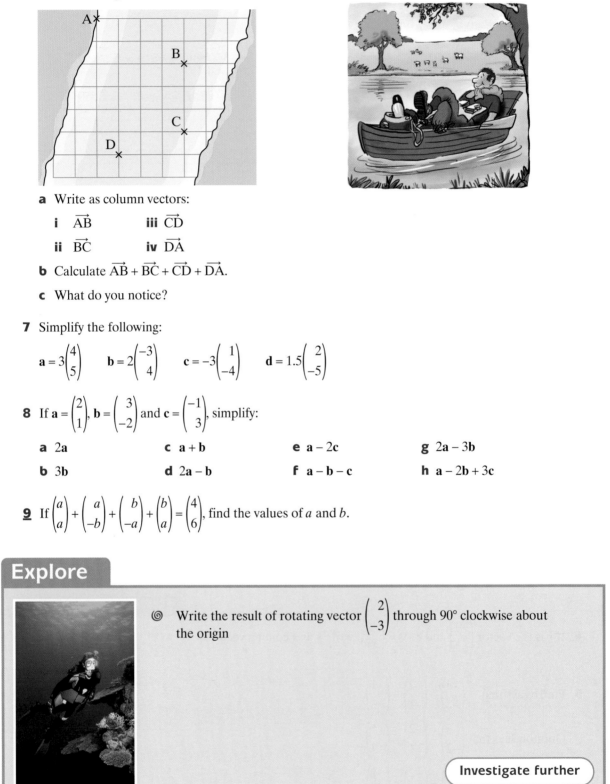

a Write as column vectors:

  i $\overrightarrow{AB}$ **iii** $\overrightarrow{CD}$

  ii $\overrightarrow{BC}$ **iv** $\overrightarrow{DA}$

b Calculate $\overrightarrow{AB} + \overrightarrow{BC} + \overrightarrow{CD} + \overrightarrow{DA}$.

c What do you notice?

7 Simplify the following:

$$\mathbf{a} = 3\begin{pmatrix} 4 \\ 5 \end{pmatrix} \qquad \mathbf{b} = 2\begin{pmatrix} -3 \\ 4 \end{pmatrix} \qquad \mathbf{c} = -3\begin{pmatrix} 1 \\ -4 \end{pmatrix} \qquad \mathbf{d} = 1.5\begin{pmatrix} 2 \\ -5 \end{pmatrix}$$

8 If $\mathbf{a} = \begin{pmatrix} 2 \\ 1 \end{pmatrix}$, $\mathbf{b} = \begin{pmatrix} 3 \\ -2 \end{pmatrix}$ and $\mathbf{c} = \begin{pmatrix} -1 \\ 3 \end{pmatrix}$, simplify:

  a $2\mathbf{a}$ **c** $\mathbf{a} + \mathbf{b}$ **e** $\mathbf{a} - 2\mathbf{c}$ **g** $2\mathbf{a} - 3\mathbf{b}$

  b $3\mathbf{b}$ **d** $2\mathbf{a} - \mathbf{b}$ **f** $\mathbf{a} - \mathbf{b} - \mathbf{c}$ **h** $\mathbf{a} - 2\mathbf{b} + 3\mathbf{c}$

9 If $\begin{pmatrix} a \\ a \end{pmatrix} + \begin{pmatrix} a \\ -b \end{pmatrix} + \begin{pmatrix} b \\ -a \end{pmatrix} + \begin{pmatrix} b \\ a \end{pmatrix} = \begin{pmatrix} 4 \\ 6 \end{pmatrix}$, find the values of a and b.

Explore

◎ Write the result of rotating vector $\begin{pmatrix} 2 \\ -3 \end{pmatrix}$ through 90° clockwise about the origin

Investigate further

Learn 2   Parallel and perpendicular vectors

Example: If $\overrightarrow{OB} = \begin{pmatrix} 3 \\ 1 \end{pmatrix}$, $\overrightarrow{OA} = \begin{pmatrix} 2 \\ 3 \end{pmatrix}$ and $\overrightarrow{OC} = \begin{pmatrix} 5 \\ -3 \end{pmatrix}$, show that A, B and C are collinear.

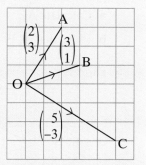

To show that points are collinear (lie on the same line), you need to show that the two lines (AB and AC) are parallel and pass through a common point.

$$\overrightarrow{AB} = -\overrightarrow{OA} + \overrightarrow{OB} = -\begin{pmatrix} 2 \\ 3 \end{pmatrix} + \begin{pmatrix} 3 \\ 1 \end{pmatrix} = \begin{pmatrix} 1 \\ -2 \end{pmatrix}$$

$$\overrightarrow{BC} = -\overrightarrow{OB} + \overrightarrow{OC} = -\begin{pmatrix} 3 \\ 1 \end{pmatrix} + \begin{pmatrix} 5 \\ -3 \end{pmatrix} = \begin{pmatrix} 2 \\ -4 \end{pmatrix} = 2\begin{pmatrix} 1 \\ -2 \end{pmatrix}$$

To get from A to B, go along $\overrightarrow{AO}$ ($= -\overrightarrow{OA}$) and $\overrightarrow{OB}$, so $AB = -\overrightarrow{OA} + \overrightarrow{OB}$

So $\overrightarrow{BC}$ is parallel to $\overrightarrow{AB}$ as $\overrightarrow{BC}$ is a multiple of $\overrightarrow{AB}$.
A, B and C are collinear as $\overrightarrow{AB}$ and $\overrightarrow{BC}$ both pass through the point B and $\overrightarrow{AB}$ is parallel to $\overrightarrow{BC}$.

Remember:

If $\mathbf{a} = \begin{pmatrix} x \\ y \end{pmatrix}$ and $\mathbf{b} = m\begin{pmatrix} x \\ y \end{pmatrix}$, then $\mathbf{a}$ and $\mathbf{b}$ are parallel

If $\mathbf{a} = \begin{pmatrix} x \\ y \end{pmatrix}$ and $\mathbf{b} = \begin{pmatrix} -x \\ -y \end{pmatrix}$, or $\mathbf{a} = \begin{pmatrix} x \\ y \end{pmatrix}$ and $\mathbf{b} = -\begin{pmatrix} x \\ y \end{pmatrix}$

then $\mathbf{a}$ and $\mathbf{b}$ are parallel, equal in length and opposite in direction

If $\mathbf{a} = \begin{pmatrix} x \\ y \end{pmatrix}$ and $\mathbf{b} = m\begin{pmatrix} -y \\ x \end{pmatrix}$, then $\mathbf{a}$ and $\mathbf{b}$ are perpendicular (see Explore on page 129)

Apply 2

1 Put the following vectors into pairs that are parallel. You can draw them if you wish.

$\mathbf{a} = \begin{pmatrix} 4 \\ -1 \end{pmatrix}$ $\mathbf{d} = \begin{pmatrix} -4 \\ -10 \end{pmatrix}$ $\mathbf{g} = \begin{pmatrix} -3 \\ 12 \end{pmatrix}$ $\mathbf{j} = \begin{pmatrix} 3 \\ -6 \end{pmatrix}$

$\mathbf{b} = \begin{pmatrix} 2 \\ 5 \end{pmatrix}$ $\mathbf{e} = \begin{pmatrix} 1 \\ -4 \end{pmatrix}$ $\mathbf{h} = \begin{pmatrix} -5 \\ -20 \end{pmatrix}$ $\mathbf{k} = \begin{pmatrix} 6 \\ -15 \end{pmatrix}$

$\mathbf{c} = \begin{pmatrix} -2 \\ 5 \end{pmatrix}$ $\mathbf{f} = \begin{pmatrix} 2 \\ 8 \end{pmatrix}$ $\mathbf{i} = \begin{pmatrix} 12 \\ -3 \end{pmatrix}$ $\mathbf{l} = \begin{pmatrix} -5 \\ 10 \end{pmatrix}$

2 If $\begin{pmatrix} x \\ 6 \end{pmatrix}$ and $\begin{pmatrix} 3 \\ 2 \end{pmatrix}$ are parallel, calculate x.

3 If $\begin{pmatrix} 6 \\ y \end{pmatrix}$, $\begin{pmatrix} x \\ -2 \end{pmatrix}$ and $\begin{pmatrix} 3 \\ 4 \end{pmatrix}$ are parallel, calculate x and y.

4 $\overrightarrow{AB} = \begin{pmatrix} 4 \\ -2 \end{pmatrix}$ and $\overrightarrow{BC} = \begin{pmatrix} -6 \\ 3 \end{pmatrix}$

    a Calculate $\overrightarrow{AC}$.

    b Prove that A, B and C are collinear.

5 Express **p**, **q**, **r** and **s** in terms of **a** and **b**.

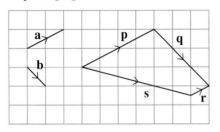

6 Norah says that **c** = 2**a**
    Oscar says **c** = −2**a**
    Phil says **a** = **b**
    Quinn says **a** + **b** + **c** = **d**

    Who is correct?
    Give reasons for your answers.

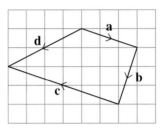

7 a If **p** = **a** + **b**, **q** = 2**a** and **r** = 2**b** − **a**, express, in terms of **a** and **b**:

    i 2**p**   **ii** **p** + **r**   **iii** **q** + 2**r**   **iv** 2**p** − **r**

    b What can you say about:

      i (**p** + **r**) and (**q** + 2**r**)   **ii** (2**p** − **r**) and **q**?

8 ACBD is a parallelogram.
    $\overrightarrow{DA}$ = **x** and $\overrightarrow{DB}$ = **y**.
    Write down, in terms of **x** and **y**:

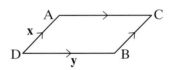

    a $\overrightarrow{DC}$

    b $\overrightarrow{AB}$

    c $\overrightarrow{DP}$, where P is the point of intersection of the two diagonals AB and CD.

9 **t** = 2**a** + 3**b**
    u = 3**a** − **b**
    v = **a** + 2**b**

    If (**t** + **u** + **v**) is parallel to (4**t** + k**v**), find the value of k.

10 $\overrightarrow{OA} = \begin{pmatrix} 2 \\ -3 \end{pmatrix}$, $\overrightarrow{OB} = \begin{pmatrix} 3 \\ -1 \end{pmatrix}$ and $\overrightarrow{OC} = \begin{pmatrix} 1 \\ 4 \end{pmatrix}$.

    M and N are the midpoints of BC and AB respectively.

    Write, as column vectors:

    a $\overrightarrow{AB}$ **c** $\overrightarrow{AC}$ **e** $\overrightarrow{OM}$ **g** $\overrightarrow{ON}$

    b $\overrightarrow{BC}$ **d** $\overrightarrow{BM}$ **f** $\overrightarrow{AN}$ **h** $\overrightarrow{MN}$

11 Write down five vectors that are perpendicular to $\begin{pmatrix} 4 \\ -3 \end{pmatrix}$.

    Draw diagrams to support your answers.

Explore

◎ The vector $\mathbf{a} = \begin{pmatrix} 2 \\ 4 \end{pmatrix}$

◎ Write a column vector parallel and equal in length to $\mathbf{a}$

◎ Write a column vector parallel and twice the length of $\mathbf{a}$

◎ Write a column vector perpendicular and equal in length to $\mathbf{a}$

◎ Write a column vector perpendicular and twice the length of $\mathbf{a}$

> **Investigate further**

Learn 3 Vector geometry

Example:

In the triangle ABC, $\overrightarrow{AB} = \mathbf{a}$ and $\overrightarrow{AC} = \mathbf{b}$.
P is the point on AB such that AP : PB = 2 : 1, and Q is the point on AC such that AQ : QC = 2 : 1.
Prove that PQ is parallel to BC, and find the ratio of the lengths PQ : BC.

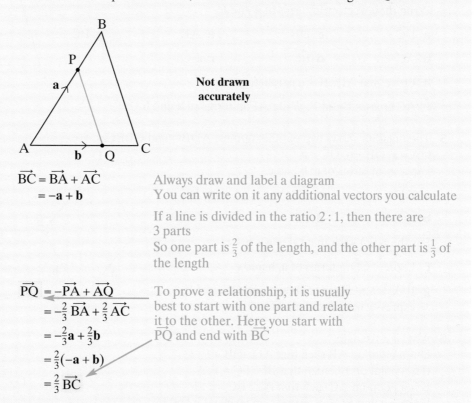

Not drawn accurately

$\overrightarrow{BC} = \overrightarrow{BA} + \overrightarrow{AC}$
$\quad = -\mathbf{a} + \mathbf{b}$

Always draw and label a diagram
You can write on it any additional vectors you calculate

If a line is divided in the ratio 2 : 1, then there are 3 parts
So one part is $\frac{2}{3}$ of the length, and the other part is $\frac{1}{3}$ of the length

$\overrightarrow{PQ} = -\overrightarrow{PA} + \overrightarrow{AQ}$
$\quad = -\frac{2}{3}\overrightarrow{BA} + \frac{2}{3}\overrightarrow{AC}$
$\quad = -\frac{2}{3}\mathbf{a} + \frac{2}{3}\mathbf{b}$
$\quad = \frac{2}{3}(-\mathbf{a} + \mathbf{b})$
$\quad = \frac{2}{3}\overrightarrow{BC}$

To prove a relationship, it is usually best to start with one part and relate it to the other. Here you start with $\overrightarrow{PQ}$ and end with $\overrightarrow{BC}$

So PQ is parallel to BC, and the ratio PQ : BC $= \frac{2}{3} : 1$ or 2 : 3

Apply 3

1 In triangle ABC, $\vec{BA}$ = **a** and $\vec{BC}$ = **b**.
   M is the midpoint of AB and N is the midpoint of BC.

   a Express, in terms of **a** and **b**:

   i $\vec{BM}$ **ii** $\vec{BN}$ **iii** $\vec{MN}$ **iv** $\vec{AC}$

   b What can you say about MN and AC?

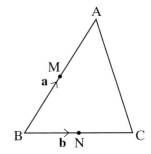

2 If $\vec{MN}$ = **a** and $\vec{MP}$ = **b**, and S is the midpoint of NP,
   prove that $\vec{MS} = \frac{1}{2}(\mathbf{a} + \mathbf{b})$.
   Use a diagram to explain your answer.

3 The vector $\vec{AB}$ = **a** and the vector $\vec{AC}$ = **b**
   Mary says that $\vec{BC}$ = **a** + **b**
   Joe says $\vec{BC}$ = **a** − **b**
   Harry says $\vec{BC}$ = **b** − **a**
   Who is right? Use a diagram to explain your answer.

4 In the diagram, ABCD is a trapezium with DC twice the length of AB.
   ABED is a parallelogram.
   $\vec{AB}$ = **a**, $\vec{AD}$ = **b**
   Write these vectors in terms of **a** and **b**:

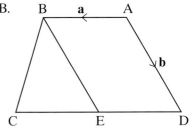

   a $\vec{DC}$ **c** $\vec{DB}$ **e** $\vec{AE}$ **g** $\vec{BC}$

   b $\vec{AC}$ **d** $\vec{BE}$ **f** $\vec{EC}$

   h Prove that AECB is a parallelogram.

   i AC crosses BE at F. Write $\vec{AF}$ in terms of **a** and **b**.

5 The diagram shows two parallelograms, ADBO and OBEC.
   $\vec{OA}$ = **a**, $\vec{OB}$ = **b**, $\vec{OC}$ = **c**
   Express, in terms of **a**, **b** and **c**:

   a $\vec{OD}$ **c** $\vec{AB}$ **e** $\vec{DC}$

   b $\vec{BC}$ **d** $\vec{BE}$

   F is the point such that $\vec{OF}$ = **a** + **b** + **c**

   f Express $\vec{CF}$ in terms of **a**, **b** and **c**

   g If A, B and E are collinear, and BE = 2AB,
        express **c** in terms of **a** and **b**.

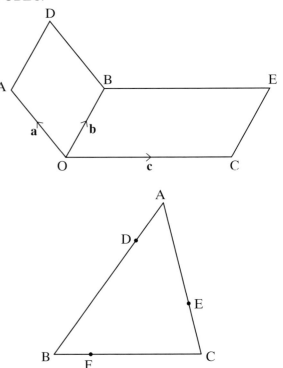

6 ABC is a triangle.
   D is a point on AB such that AD : DB = 1 : 3
   F is a point on BC such that BF : FC = 1 : 3
   E is a point on AC such that AE : EC = 2 : 1
   If $\vec{BA}$ = **a** and $\vec{BC}$ = **b**, express, in terms of **a** and **b**:

   a $\vec{BF}$ **b** $\vec{BD}$ **c** $\vec{FD}$ **d** $\vec{BE}$ **e** $\vec{DE}$

7 Vector $\overrightarrow{EA} = \mathbf{a} = \begin{pmatrix} 2 \\ 4 \end{pmatrix}$, $\overrightarrow{EB} = \mathbf{b} = \begin{pmatrix} 1 \\ 5 \end{pmatrix}$ and $\overrightarrow{EC} = \mathbf{c} = \begin{pmatrix} 6 \\ y \end{pmatrix}$

If A, B and C are collinear, find the value of y.

8 ABEF is a parallelogram. D is the midpoint of BE and C is the midpoint of AF.

M is the point on BC such that BM : MC = 2 : 1

If $\overrightarrow{AB} = \mathbf{b}$ and $\overrightarrow{AC} = \mathbf{c}$, prove that $\overrightarrow{AE} = 3\,\overrightarrow{AM}$

Use a diagram to explain your answer.

9 In the triangle XYZ, M is the midpoint of YZ, and S is the midpoint of XZ.

YS crosses XM at R such that XR : RM = 2 : 1

$\overrightarrow{XY} = \mathbf{a}$ and $\overrightarrow{XZ} = \mathbf{b}$.

Express, in terms of **a** and **b**:

a $\overrightarrow{YS}$ **c** $\overrightarrow{ZM}$ **e** $\overrightarrow{XR}$ **g** $\overrightarrow{RS}$

b $\overrightarrow{ZY}$ **d** $\overrightarrow{XM}$ **f** $\overrightarrow{XS}$

h Use your answers to **a** and **g** to calculate the ratio YS : RS.

Explore

◎ ABCD is a quadrilateral whose diagonals bisect each other at M

◎ $\overrightarrow{AB} = \mathbf{a}$ and $\overrightarrow{BC} = \mathbf{b}$

◎ Find $\overrightarrow{AC}$, $\overrightarrow{AM}$, $\overrightarrow{MC}$, $\overrightarrow{BM}$, $\overrightarrow{MD}$, $\overrightarrow{AD}$ and $\overrightarrow{CD}$ in terms of **a** and **b**

◎ What can you say about ABCD?

Investigate further

Vectors

ASSESS

The following exercise tests your understanding of this chapter, with the questions appearing in order of increasing difficulty.

1 a Write down the coordinates of the images of each point after the vector translations shown.

Object	Vector	Object	Vector
i (3, 5)	$\begin{pmatrix} 4 \\ 3 \end{pmatrix}$	**iii** (−5, −8)	$\begin{pmatrix} -4 \\ 4 \end{pmatrix}$
ii (−2, 7)	$\begin{pmatrix} 4 \\ -3 \end{pmatrix}$		

b Write down the vector translation that transforms:

    i $(4, 1)$ into $(6, -3)$ **ii** $(3, -3)$ into $(1, -2)$ **iii** $(-4, -8)$ into $(-3, -9)$?

2 Write down the vectors **a**, **b**, **c**, **d**, **e**, and **f**.

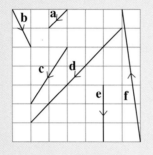

3 $\mathbf{m} = \begin{pmatrix} 2 \\ 4 \end{pmatrix}$ and $\mathbf{n} = \begin{pmatrix} -6 \\ 3 \end{pmatrix}$

Write as column vectors:

    i $2\mathbf{m}$ **ii** $-3\mathbf{n}$ **iii** $2\mathbf{m} - 3\mathbf{n}$ **iv** $\frac{1}{2}\mathbf{n} + \frac{1}{4}\mathbf{m}$

4 Draw a triangle PQR and mark $\overrightarrow{PQ}$ as vector **a** and $\overrightarrow{PR}$ as vector **b**.
Mark in the midpoints of PQ and QR as A and B respectively.

  a Write down the following vectors, in terms of **a** and **b**.

    i $\overrightarrow{QR}$ **ii** $\overrightarrow{QB}$ **iii** $\overrightarrow{AQ}$ **iv** $\overrightarrow{AB}$

  b What conclusions can you draw when comparing vectors $\overrightarrow{AB}$ and $\overrightarrow{PR}$?

Try a real past exam question to test your knowledge:

5 ABCDEF is a regular hexagon with centre O.

$\overrightarrow{OA} = \mathbf{a}$ and $\overrightarrow{AB} = \mathbf{b}$.

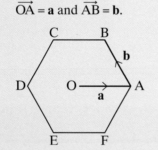

  a Find expressions, in terms of **a** and **b**, for:

    i $\overrightarrow{OB}$ **ii** $\overrightarrow{AC}$ **iii** $\overrightarrow{EC}$

  b The positions of points P and Q are given by the vectors:

    $\overrightarrow{OP} = \mathbf{a} - \mathbf{b}$ $\overrightarrow{OQ} = \mathbf{a} + 2\mathbf{b}$

    i Draw and label the positions of points P and Q on the diagram.

    ii Hence, or otherwise, deduce an expression for $\overrightarrow{PQ}$.

Spec A, Higher Paper 1, June 04

10 Graphs of linear functions

What you should already know ...

■ Plot coordinates in all four quadrants

■ Plot graphs of linear functions

■ Discuss and interpret graphs of real-life situations

VOCABULARY

Gradient – a measure of how steep a line is

$$\text{Gradient} = \frac{\text{change in vertical distance}}{\text{change in horizontal distance}} = \frac{y}{x}$$

positive gradient   negative gradient

Linear graph – the graph of a linear function of the form $y = mx + c$; if c is zero, the graph is a straight line through the origin (the point $(0, 0)$) indicating that y is directly proportional to x; if m is zero, the graph is parallel to the x-axis

Intercept – the y-coordinate of the point at which the line crosses the y-axis

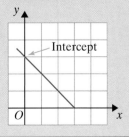

Learn 1 Drawing straight-line graphs

Examples:

a Draw the lines $x = 3$ and $y = -4$ and find where they meet.

b Draw the straight-line graph $y = 2x - 3$ for values of x from -2 to 4.

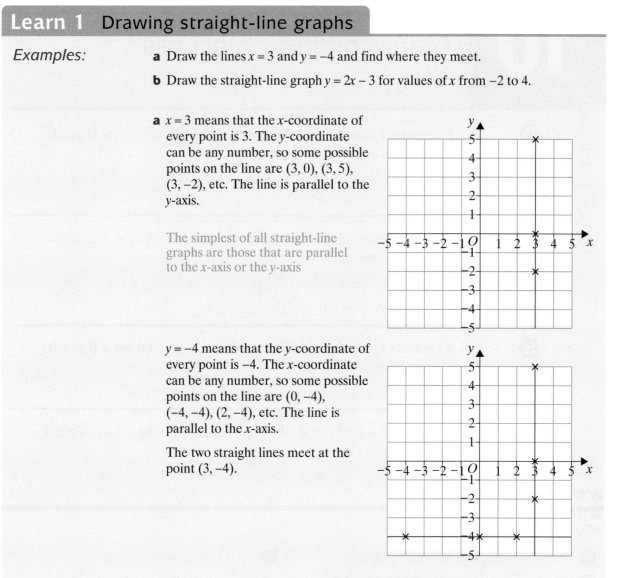

a $x = 3$ means that the x-coordinate of every point is 3. The y-coordinate can be any number, so some possible points on the line are $(3, 0)$, $(3, 5)$, $(3, -2)$, etc. The line is parallel to the y-axis.

The simplest of all straight-line graphs are those that are parallel to the x-axis or the y-axis

$y = -4$ means that the y-coordinate of every point is -4. The x-coordinate can be any number, so some possible points on the line are $(0, -4)$, $(-4, -4)$, $(2, -4)$, etc. The line is parallel to the x-axis.

The two straight lines meet at the point $(3, -4)$.

b Use the equation $y = 2x - 3$ to work out some points on the line.

The equation says that you double the x-coordinate and subtract 3 to find the y-coordinate, so choose any value of x from -2 to 4 and work out the corresponding value of y

If x is 0, $y = 2 \times 0 - 3 = 0 - 3 = -3$
So one point on the graph is $(0, -3)$.

It is possible to work out many more pairs of values but three points are enough for a straight line

If x is 4, $y = 2 \times 4 - 3 = 8 - 3 = 5$
So another point on the graph is $(4, 5)$.

If x is -2, $y = 2 \times (-2) - 3 = -4 - 3 = -7$
So a third point on the line is $(-2, -7)$.

For each value of x there is just one value of y that fits the equation

Each pair of (x, y) values is a point on the line

There is an infinite number of points on any straight line

Now mark at least three of the points on squared paper and join them with a straight line.

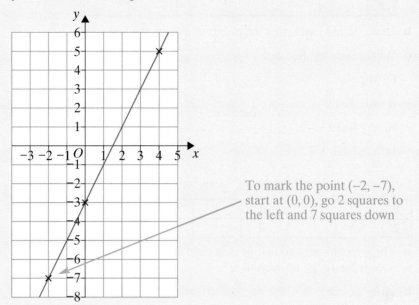

To mark the point $(-2, -7)$, start at $(0, 0)$, go 2 squares to the left and 7 squares down

Other coordinates can be found from a table, for example, for the graph $y = 2x - 3$:

x	−2	−1	0	1	2	3	4
y	−7	−5	−3	−1	1	3	5

Apply 1

1 Draw an x-axis and a y-axis, each going from −6 to 6.

   a On the axes, draw the lines $x = 2$ and $x = -4$.   These lines are parallel to the y-axis

   b Draw the lines $y = 5$ and $y = -2$.   These lines are parallel to the x-axis

   c Where does the line $x = 2$ meet the line $y = -2$?

   d Where does the line $x = -4$ meet the line $y = 5$?

   e How can you work out the points of intersection without using the graph?

2 Here is a table of values for the line $y = 2x$

x	−3	−2	−1	0	1	2	3
y	−6	−4	−2	0	2	4	6

   a Draw an x-axis and a y-axis, the x-axis going from −3 to 3 and the y-axis going from −6 to 6.
      Plot the points in the table of values and join them with a straight line.

   b Write down the coordinates of the point on the line:

      i with an x-coordinate of 0.5

      ii with a y-coordinate of 5.

3 a Copy and complete this table of x- and y-values for the equation $y = x + 3$

x	−4	−3	−2	−1	0	1
y	−1		1			4

b Draw suitable axes, plot the points and draw the line $y = x + 3$

c Where does the line cut:

  i the x-axis

  ii the y-axis

  iii the line $x = -3$

  iv the line $y = 2$?

4 Copy and complete this table of values for the equation $y = 3x - 1$

x	−2	−1	0	1	2	3
y	−7		−1			8

a The x-coordinates go up one unit each time. How many units do the y-coordinates go up each time?

b Draw an x-axis labelled from −2 to 3 and a y-axis labelled from −7 to 8 and plot the points.
Draw a straight line through the points.

c Find the y-coordinate of a point on the line with x-coordinate 1.5

d Find the x-coordinate of a point on the line with a y-coordinate of 4.

5 a Make a table of values for $y = 2x + 4$ for values of x from −2 to 3.

b Draw suitable axes, plot the points and draw the line $y = 2x + 4$

c Write the coordinates of the point where $y = 2x + 4$ crosses:

  i the x-axis

  ii the y-axis

  iii the line $x = -1$

  iv the line $y = 3$

6 a For the line $y = 3x + 1$, copy and complete these coordinate pairs:
  $(-3, ...)\ (0, ...)\ (3, ...)$

b Draw suitable axes on squared paper, mark the three points and join them with a straight line.

c Write down the coordinates of the points where the line $y = 3x + 1$ crosses the x-axis and the y-axis.

d Use the equation to work out the coordinates of the points where the line crosses the x- and y-axes.

7 Draw an x-axis and a y-axis, each going from −10 to 10.

a On these axes, draw the lines:

  i $y = 2x + 4$ **ii** $y = \frac{1}{2}x$

b i Where does the line $y = \frac{1}{2}x$ cut the axes?

  ii How can you find this from the equation?

8 Which of these points lie on the line $y = \frac{1}{3}x$?

$(3, 1), (3, 0), (0, 0), (-3, -1), (1, 3), (2, 6), (0.9, 0.3)$

Show how you found your answers.

9 a Draw the straight line that goes through the points $(-1, -4)$, $(0, 0)$ and $(2, 8)$.

b Which of these is the equation of the line?

$y = x + 6 \qquad y = x - 3 \qquad y = 4x$

10 The equation $x + y = 6$ means that the x-coordinates and y-coordinates add up to 6. So one point on this line is $(5, 1)$.

a Find at least two other points on the line.

b Draw an x-axis and a y-axis, each going from -10 to 10.
Plot the points and join them with a straight line.

c On the same diagram, draw the line $x + y = 8$

d What is the same and what is different about the two lines?

<u>11</u> The diagram shows several straight-line graphs.

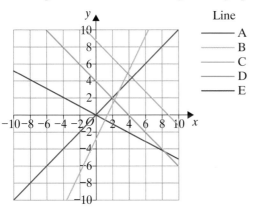

Line
——— A
——— B
——— C
——— D
——— E

a Write down the letters of two lines that are parallel.

b Write down the letters of two lines that are perpendicular.

c Write down the letters of the lines that go through the origin.

d Write down the letter of the line with the greatest slope.

e Write down the coordinates of three points on Line A.

Explore

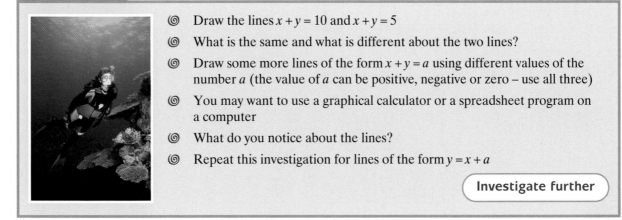

◎ Draw the lines $x + y = 10$ and $x + y = 5$

◎ What is the same and what is different about the two lines?

◎ Draw some more lines of the form $x + y = a$ using different values of the number a (the value of a can be positive, negative or zero – use all three)

◎ You may want to use a graphical calculator or a spreadsheet program on a computer

◎ What do you notice about the lines?

◎ Repeat this investigation for lines of the form $y = x + a$

Investigate further

Learn 2 Recognising the equations of straight-line graphs and finding their gradients

Examples:

a Find the gradient of the line with equation:

   i $y = 4x + 2$ **ii** $y = 3x + 6$

b Rearrange the following equations in the form $y = mx + c$

   i $x + y = 4$ **ii** $2y + 3x = 12$

a i First think about the line with equation $y = 4x$
   Here is the table of values:

> The x-values go up in ones and the y-values go up in fours

x	−3	−2	−1	0	1	2
y = 4x	−12	−8	−4	0	4	8

Now back to the line with equation $y = 4x + 2$

Here is the table of values:

x	−3	−2	−1	0	1	2
y = 4x	−12	−8	−4	0	4	8
y = 4x + 2	−10	−6	−2	2	6	10

To find the y-values for $y = 4x + 2$,
add 2 to the y values for $y = 4x$

The y-values still go up in fours

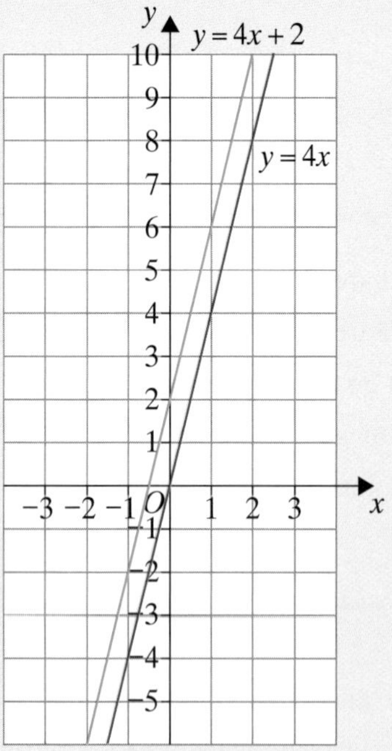

$y = 4x + 2$

$y = 4x$

The diagram shows the lines $y = 4x$ and $y = 4x + 2$
The lines are parallel; they both have the same gradient.

Adding 2 to make $y = 4x + 2$ does not change the gradient but moves the line up 2 squares

The gradient of each line is 4 as each line goes up the page 4 units for every unit across the page (Count the squares)

$y = 4x + 2$ ← y-intercept is at +2 so passing through (0, +2)

↑ gradient

ii For the line with equation $y = 3x - 6$

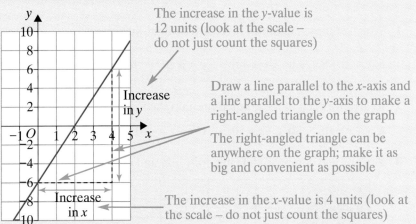

The increase in the y-value is 12 units (look at the scale – do not just count the squares)

Draw a line parallel to the x-axis and a line parallel to the y-axis to make a right-angled triangle on the graph

The right-angled triangle can be anywhere on the graph; make it as big and convenient as possible

The increase in the x-value is 4 units (look at the scale – do not just count the squares)

The gradient of the graph is how many units it goes up for every unit across.

So gradient $= \dfrac{\text{increase in } y}{\text{increase in } x} = \dfrac{12}{4} = 3$

The graph goes up 3 units for every unit across.
It crosses the y-axis at -6, as the equation of this graph is $y = 3x - 6$

$y = 3x - 6$ ← y-intercept is at -6 so passing through $(0, -6)$

gradient

Any graph parallel to $y = 3x - 6$ has gradient 3 and any graph with gradient 3 is parallel to $y = 3x - 6$

Any equation that can be written in the form $y = mx + c$ is a linear equation and can be drawn as a straight-line graph.

b i

$x + y = 4$
$y = 4 - x$ — Subtract x from both sides
$y = -x + 4$ — Rearranging in the form $y = mx + c$

This is a straight line with gradient $= -1$
and y-intercept at 4 so passing through the point $(0, 4)$

A gradient of -1 slopes from top left to bottom right

ii $2y + 3x = 12$

$2y = 12x - 3x$ — Subtract $3x$ from both sides
$y = 6 - \frac{3}{2}x$ — Divide each side by 2
$y = \frac{3}{2}x + 6$ — Rearranging in the form $y = mx + c$

This is a straight line with gradient $= -\frac{3}{2}$
and y-intercept at 6 so passing through the point $(0, 6)$.

A gradient of $-\frac{3}{2}$ slopes from top left to bottom right

139

Apply 2

1 Which of these equations represent straight-line graphs?

a $y = 2x + 8$ **b** $2y = x + 8$ **c** $x = 2y$ **d** $y = \dfrac{x}{2}$ **e** $y = \dfrac{2}{x}$

2 Rearrange each of these equations into the form $y = mx + c$:

a $y + x = 10$ **b** $y - x = 10$ **c** $y + x = -10$ **d** $y - x = -10$

3 a Copy and complete the table for $y = 2x$ and $y = 2x - 3$

x	−2	−1	0	1	2
y = 2x			0		
y = 2x − 3			−3		

   b Draw an x-axis and a y-axis, the x-axis going from −2 to 2 and the y-axis going from −8 to 4. On the axes, plot each set of points from the table and join them with a straight line.

   c Use your diagram to find the gradient of each line.

4 Repeat question **3** for the lines $y = 5x$ and $y = 5x - 4$
(The y-axis will need to go from −15 to 10.)

5 a Draw these straight lines and work out their gradients.

     i $y = 2x$ **ii** $y = \frac{1}{2}x$ **iii** $y = -2x$ **iv** $y = -\frac{1}{2}x$

   b Use the straight-line graphs in part **a** to help you to work out the gradients of these lines:

     i $y = 2x + 5$ **ii** $y = \frac{1}{2}x + 5$ **iii** $y = -2x + 5$ **iv** $y = -\frac{1}{2}x + 5$

   c Work out the gradients of these lines without drawing the graphs.

     i $y = \frac{1}{3}x$, $y = -\frac{1}{3}x$, $y = \frac{1}{3}x - \frac{2}{3}$, $y = -\frac{1}{3}x - \frac{2}{3}$

     ii $y = -0.2x$, $y = 0.2x - 0.5$, $y = 0.2x$, $y = -0.2x - 0.5$

6 Write down the gradients of the straight-line graphs in this diagram.

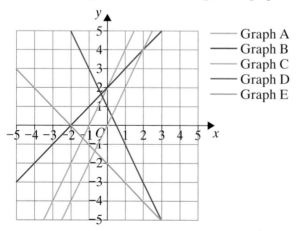

Graph A
Graph B
Graph C
Graph D
Graph E

7 a Rearrange each of these equations into the form $y = mx + c$:

i $2y + x = 6$ **iii** $2y - x = -6$ **v** $3x + 4y = 8$ **vii** $-3x + 4y = -8$

ii $y - 2x = 6$ **iv** $y + 2x = -6$ **vi** $3x - 4y = 8$ **viii** $3x + 4y + 8 = 0$

b Find the gradients of the lines.

8 The diagram shows the line $y = 3x - 5$
AB is 2 units long.
How long is BC?

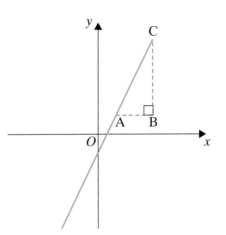

9 Get Real!

The equation $y = \frac{9}{5}x + 32$ gives a straight-line graph that converts
Celsius temperatures (x) to Fahrenheit temperatures (y).

a What is the gradient of the line?

b What is its y-intercept?

c Interpret the meaning of these numbers.

d Complete this table of values for the line $y = \frac{9}{5}x + 32$

Temperature (degrees Celsius)		20	100
Temperature (degrees Fahrenheit)	32		

e Draw the line, choosing scales so that the three points in
the table fit on the diagram.

f Use the line to convert the temperature 35°C to
Fahrenheit.

HINT The y-intercept of a straight
line is the y-coordinate of the point
where the line cuts the y-axis.

Explore

◎ The general form of straight-line graphs is
$y = mx + c$

All straight lines, except those
parallel to the y-axis, can be
expressed in this form

◎ Choose simple numbers for m and c and explore how different values affect
the lines. (You may find it useful to use a graphical calculator or a graph-
drawing computer program)

Equations of graphs parallel to
the y-axis are of the form $x = a$

Investigate further

Learn 3   Parallel and perpendicular straight-line graphs

Examples:

a Find the gradient of the straight line $2y + x = 5$ and find the equation of the line parallel to $2y + x = 5$ that cuts the y-axis at $(0, 4)$.

Instead of doing this calculation, it is easy to find the gradient of a straight line by looking at its equation. This equation is not in the form $y = mx + c$, so you must first rearrange it as in Learn 2:

When the equation is in the form $y = mx + c$, m is the gradient, so in this example the gradient is $-\frac{1}{2}$

$$2y + x = 5$$
$$2y = -x + 5 \qquad \text{Subtract } x \text{ from each side}$$
$$y = -\frac{1}{2}x + \frac{5}{2} \qquad \text{Divide both sides by 2}$$

Any line parallel to $2y + x = 5$ will also have a gradient of $-\frac{1}{2}$, so the equation of a parallel line will be of the form $y = -\frac{1}{2}x + c$

In the equation $y = mx + c$, c is the y-coordinate of the point where the line cuts the y-axis. This coordinate is also called the y-intercept.

So, the line that is parallel to $2y + x = 5$ and goes through the point $(0, 4)$ has the equation $y = -\frac{1}{2}x + 4$

You can also be write this in other forms such as $2y + x = 8$

b The diagram shows the line $y = 3x - 6$ together with a line that is perpendicular to it. Show that the product of the gradients of the two lines is -1 and that the perpendicular line has the equation $y = -\frac{1}{3}x + 4$

Right-angled triangles show the x and y increases for both lines.

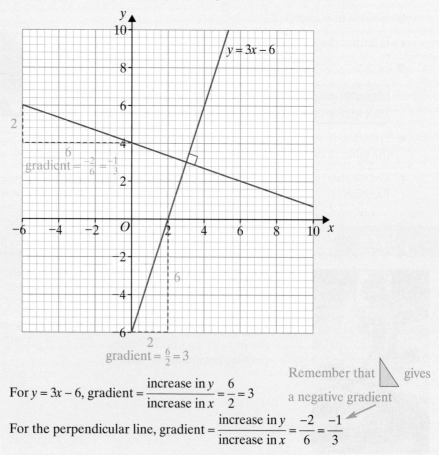

For $y = 3x - 6$, gradient $= \dfrac{\text{increase in } y}{\text{increase in } x} = \dfrac{6}{2} = 3$

For the perpendicular line, gradient $= \dfrac{\text{increase in } y}{\text{increase in } x} = \dfrac{-2}{6} = \dfrac{-1}{3}$

Remember that ◺ gives a negative gradient

When two lines are perpendicular, one will have a positive gradient and the other will have a negative gradient.

Lines parallel to the axes are the exception, as one has a zero gradient and one has an infinite gradient

The product of the two gradients is $3 \times (-\frac{1}{3}) = -1$

This example is a demonstration of the general rule that the product of the gradients of any two perpendicular lines is -1.

Apply 3

1 Find the equations of these lines:

   a Gradient of 3;   goes through $(0, 0)$

   b Gradient of 3;   goes through $(0, 6)$

   c Gradient of 3;   goes through $(0, -2)$

   d Gradient of -4;   goes through $(0, 0)$

   e Gradient of -4;   goes through $(0, -2)$

   f Gradient of -2;   goes through $(0, 4)$

   g Gradient of -1;   goes through $(0, 8)$

   h Gradient of $\frac{1}{3}$;   goes through $(0, -2)$

   i Gradient of -1.3;   goes through $(0, 0.3)$

   j Gradient of -1;   goes through $(0, 0)$

2 Find the gradient of the straight line that goes through each of the following pairs of points.

   a $(0, 0)$ and $(5, 10)$

   b $(0, 0)$ and $(10, 5)$

   c $(0, 8)$ and $(8, 0)$

   d $(2, 4)$ and $(6, 6)$

   e $(3, 5)$ and $(-1, -3)$

3 Find the equations of the lines in question **2**.

4 The diagram shows some parallel and perpendicular straight-line graphs.

  a Use the diagram to work out the gradient of each line.

  b Identify the pairs of parallel lines in the diagram and show that their gradients are the same.

  c Identify the pairs of perpendicular lines and show that the products of their gradients are −1.

  d Find the equation of each line.

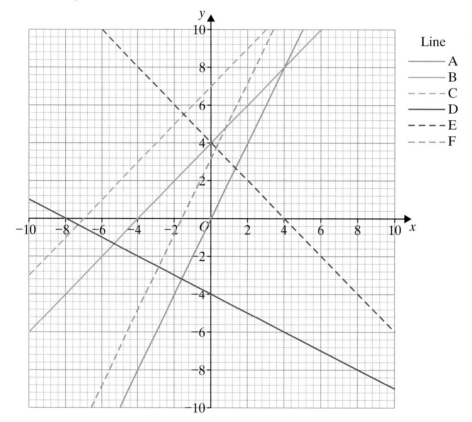

5 **a** Is it correct to say that the lines $2y + x = 3$ and $y = -\frac{1}{2}x$ are parallel? Give a reason for your answer.

  b Is it correct to say that the lines $y + 2x = 3$ and $y = -\frac{1}{2}x$ are perpendicular? Give a reason for your answer.

6 Find the three sets of parallel lines in these equations.

$y = 2x$	$y = 5 + 3x$	$y + x = 10$	$x = \frac{1}{2}y$
$y - 2x = 3$	$3x - y + 2 = 0$	$x = \frac{1}{2}y + 1$	$y = 3x - 8$
$y + x = 0.5$	$2y - 4x = 3$	$3x - y + 2 = 0$	$y = -x - 3$
$2x + 2y - 1 = 0$	$\frac{1}{3}y = x + 1$	$y = 5 - x$	$y = -x$

7 Here are the equations of three straight-line graphs.
For each pair of graphs, decide whether they are
parallel, perpendicular or neither.

$3x + y = 6$, $y = 3x + 6$, $2y = 6x + 5$

Show how you decided.

8 On graph paper:

a draw the line $3x + 4y = 12$

b i draw a line parallel to $3x + 4y = 12$ that passes
through the point $(4, 2)$

ii find the equation of this line

c i draw a line perpendicular to $3x + 4y = 12$ that passes through the
point $(4, 2)$

ii find the equation of this line.

> **HINT** For equations given in this form,
> find two points like this:
> - put $x = 0$ and find the corresponding y-value
> - put $y = 0$ and find the corresponding x-value

9 Find the three pairs of perpendicular lines in these equations.

$y = 1 - 2x$ $\quad\quad y = \dfrac{-x}{4}$ $\quad\quad y + x = 5$

$y = 4x + 3$ $\quad\quad 2y - x = 5$ $\quad\quad y = x + 5$

10 Find the three pairs of perpendicular lines in these equations.

$3y = 4x + 18$ $\quad\quad 4y + 3x = 8$ $\quad\quad 3y = 2x + 3$

$4y + 2x = 6$ $\quad\quad y + 2x = 6$ $\quad\quad 4y = 2x + 6$

$4y = 3x + 18$ $\quad\quad 3x + 2y = 4$ $\quad\quad 2x + 3y = 5$

11 Find the equation of the perpendicular bisector of the line segment
joining each pair of points.

a $(0, 0)$ and $(10, 10)$ $\quad\quad$ **d** $(5, 6)$ and $(5, -6)$ $\quad\quad$ **g** $(-2, 5)$ and $(2, 5)$

b $(10, 0)$ and $(0, 10)$ $\quad\quad$ **e** $(5, 6)$ and $(5, -10)$

c $(4, 0)$ and $(0, 4)$ $\quad\quad$ **f** $(-2, 5)$ and $(4, 5)$

12 Find in the form $y = mx + c$ or $x = a$ the equation of the line passing
through each of the pairs of points in question **11**.

13 A triangle has its vertices at the points $(-2, 0)$, $(5, 1)$ and $(-1, 3)$.
Find whether the triangle is right-angled or not, showing your working.

14 The equation $y = ax + b$, where a and b are constants, is the equation of a straight-line graph.

Copy and complete the table to show whether the equations represent graphs parallel to $y = ax + b$, perpendicular to $y = ax + b$ or neither.

Equation of line	Parallel to $y = ax + b$	Perpendicular to $y = ax + b$	Neither parallel nor perpendicular
$y = ax$			
$y = -ax$			
$y = ax - 3$			
$y + ax = b$			
$y - ax = 2b$			
$y = 1 - \frac{x}{a}$			
$ay + x = 4$			
$y = \frac{x}{a} + 1$			

15 Show why the product of the gradients of two perpendicular lines is always -1.

Explore

- ◎ On graph paper, draw a rhombus with its vertices on grid-points
- ◎ Calculate the gradients of the sides of the rhombus and show that opposite sides are parallel
- ◎ Show that the diagonals of the rhombus are the perpendicular bisectors of each other

Investigate further

Graphs of linear functions

ASSESS

The following exercise tests your understanding of this chapter, with the questions appearing in order of increasing difficulty.

1 a Write down the gradient of each of the following straight lines:

  i $y = 5x + 2$ **iii** $y = 7x - 1$ **v** $4x + 2y = 5$

  ii $y = 8 + x$ **iv** $y = 5 - 3x$ **vi** $5x - 2y = 6$

  b i Nadia says that the line $y = 7x - 1$ crosses the y-axis at $(-1, 0)$. Is this correct? Give a reason for your answer.

  ii Orla says that the line $y = 5 - 3x$ has the same gradient as the line $3x + y = 8$. Is this correct? Give a reason for your answer.

2 Shane sells luxury electrical goods.
He is paid a basic wage each month plus a percentage commission
on his sales that month.
If his sales for the month are £20 000 he is paid £1400.
If his sales are £50 000 he is paid £2300.

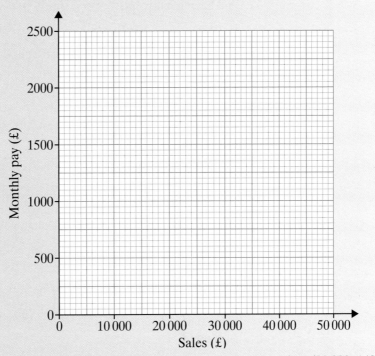

a Use a grid like the one shown above to plot the points (20 000, 1400)
and (50 000, 2300).

Draw a straight-line graph through your points and use it to answer
the following questions.

b i How much is Shane's basic wage?

ii What is the gradient of the graph?

iii What is the percentage rate of his commission?

3 a Write down the equations of the straight lines parallel to $y = 3x - 2$ and
having intercepts:

i $(0, 2)$ **ii** $(0, -3)$ **iii** $(0, -25)$ **iv** $(0, \frac{1}{2})$

b Write down the equations of the straight lines parallel to $y = 7 - \frac{1}{8}x$ and
having intercepts:

i $(0, 2)$ **ii** $(0, -3)$ **iii** $(0, -25)$ **iv** $(0, \frac{1}{2})$

c Write down the equations of the straight lines parallel to $3x - 5y + 9 = 0$
and having intercepts:

i $(0, 2)$ **ii** $(0, -3)$ **iii** $(0, -25)$ **iv** $(0, \frac{1}{2})$

4 a Write down the equations of the straight lines perpendicular to $y = 3x - 2$ and having intercepts:

    i $(0, 4)$ **ii** $(0, -2)$ **iii** $(0, -30)$ **iv** $(0, 1\frac{1}{2})$

  b Write down the equations of the straight lines perpendicular to $y = 7 - \frac{1}{8}x$ and having intercepts:

    i $(0, 4)$ **ii** $(0, -2)$ **iii** $(0, -30)$ **iv** $(0, 1\frac{1}{2})$

  c Write down the equations of the straight lines perpendicular to $3x - 6y + 8 = 0$ and having intercepts:

    i $(0, 4)$ **ii** $(0, -2)$ **iii** $(0, -30)$ **iv** $(0, 1\frac{1}{2})$

5 a The line joining points A$(-1, 4)$ and B$(3, 8)$ is a chord of a circle.

    i Find the gradient of AB.

    ii Find the midpoint of AB.

    iii Hence find the equation of the perpendicular bisector of this chord.

  b C is the point $(7, 4)$ and BC is another chord of the circle.
Find the equation of the perpendicular bisector of BC.

  c Use these equations to find the centre of the circle, O.
(The perpendicular bisector of any chord passes through the centre of the circle.)

  d Plot the points A, B, C and O on a graph.
Show that a circle can be drawn, with O as centre, which passes through A, B and C.

11 Similarity and congruence

C ▶ **Examiners would normally expect students who get a C grade to be able to:**

Match one side and one angle of congruent triangles, given some dimensions

B ▶ **Examiners would normally expect students who get a B grade also to be able to:**

Match sides and angles of similar triangles, given some dimensions

A ▶ **Examiners would normally expect students who get an A grade also to be able to:**

Prove that two triangles are congruent

Prove the construction theorems

Find the area of a 2-D shape, given the area of a similar shape and the ratio

Find the volume of a 3-D solid, given the volume of a similar solid and the ratio

What you should already know ...

- Simplify ratios and use ratios to set up equations
- Use enlargements and scale factors
- Solve equations of the form $\frac{x}{10} = \frac{9}{15}$
- Opposite angles and alternate angles

- Square and cube numbers
- Work with numbers in surd form
- Construct bisectors of lines and angles

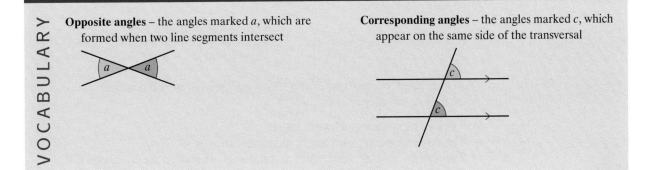

Opposite angles – the angles marked a, which are formed when two line segments intersect

Corresponding angles – the angles marked c, which appear on the same side of the transversal

Alternate angles – the angles marked a, which appear on opposite sides of the transversal

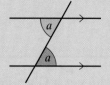

Ratio – the ratio of two or more numbers or quantities is a way of comparing their sizes, for example, if a school has 25 teachers and 500 students, the ratio of teachers to students is 25 to 500, or 25 : 500 (read as 25 to 500)

Similar – shapes are similar if their corresponding angles are equal *and* their corresponding sides are in the same ratio.

Congruent – exactly the same size and shape; one of the shapes might be rotated or flipped over

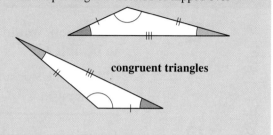

congruent triangles

Learn 1 Similarity of triangles and other 2-D shapes

Example:

Triangles ABC and DFE are similar.
State the ratio of the corresponding sides and find x.

The order of the letters is important. Triangle ABC is similar to triangle DFE. This means that angle A = angle D, angle B = angle F and angle C = angle E.
Also, AB corresponds to DF, BC corresponds to FE and AC corresponds to DE

The triangles are similar because the corresponding angles are equal

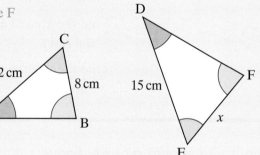

Since the triangles are similar, the corresponding sides are in the same ratio.
AC : DE = 12 : 15 = 4 : 5 so the ratio of the corresponding sides is 4 : 5.
So the ratio BC : FE = the ratio AC : DE = 4 : 5

$$\text{and } \frac{BC}{FE} = \frac{AC}{DE}$$

Form an equation in x and solve it to find x; it is easier if you keep x on the top of the fraction

$$\frac{8}{x} = \frac{4}{5}$$

$$\frac{x}{8} = \frac{5}{4} \qquad \text{Turning upside down}$$

$$x = \frac{5}{4} \times 8 \qquad \text{Multiply both sides by 8}$$

$$x = 10 \text{ cm}$$

You only need to know one of the following facts about two triangles to be able to say they are similar:

- The corresponding angles are equal.
- The corresponding sides are in the same ratio.
- Two sides are in the same ratio and the angles between them are equal.

Apply 1 ⊞

1 These pairs of shapes are similar but are not drawn accurately.
Find the missing sides and angles.

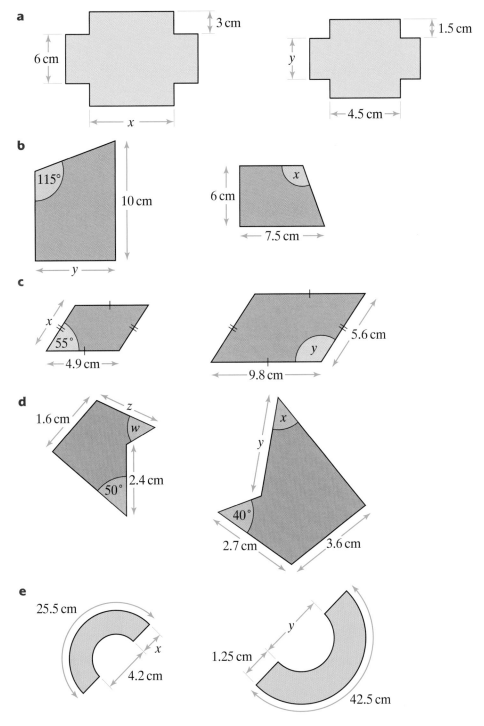

a 3 cm, 6 cm, x, 1.5 cm, y, 4.5 cm

b 115°, 10 cm, y, x, 6 cm, 7.5 cm

c x, 55°, 4.9 cm, 5.6 cm, y, 9.8 cm

d 1.6 cm, z, w, 2.4 cm, 50°, x, y, 40°, 2.7 cm, 3.6 cm

e 25.5 cm, x, 4.2 cm, y, 1.25 cm, 42.5 cm

2 These three triangles are similar. Find the missing sides.

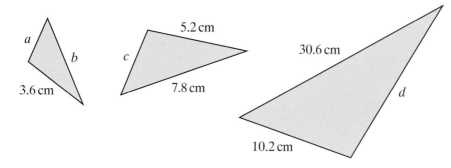

3 Triangle ABC is similar to triangle DFE.
The ratio of the corresponding sides is 4 : 7.
BC = 28 cm and DE = 42 cm.
Find the length of another side from each triangle.

4 Each pair of triangles is similar.
Write down the ratio of the corresponding sides and find the missing values.

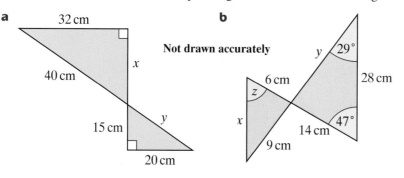

a

32 cm

b

Not drawn accurately

x

y / 29°

40 cm

6 cm

28 cm

z

15 cm

y

x

47°

20 cm

14 cm

9 cm

5 The diagram shows two 'V'-shapes. Aliya says that you need to change the width of the larger letter to 8.4 cm to make the V-shapes similar.
Chloe says that you need to change the height of the larger letter to 8.4 cm to make the V-shapes similar.
Who is right? Give a reason for your answer.

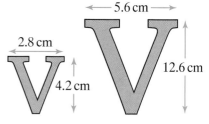

5.6 cm

2.8 cm

4.2 cm

12.6 cm

6 Get Real!
Tom took this photo of the Eiffel tower on holiday in France.
He remembers that the width of the base is 125 m, but he can't remember the height of the tower.
Use the measurements of the photo to work out the approximate height of the Eiffel tower.

7 Get Real!

This is a diagram of how a slide projector displays an image.

a What will be the height of the image?

b If the projector is moved further away from the screen what happens to the size of the image?

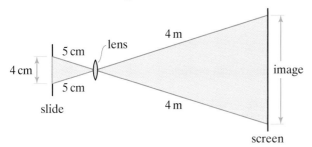

8 Get Real!

An engineer needs to construct a temporary bridge over the river from B to A. He surveys the area and draws a sketch with his measurements on it.

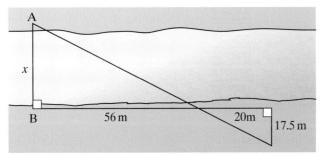

Form an equation in x and solve it to find the length of the bridge.

9 Find the sides marked with letters in these diagrams.

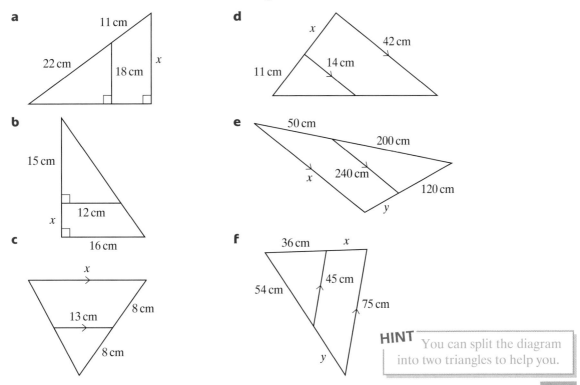

HINT You can split the diagram into two triangles to help you.

10 **Get Real!**

Sylvie wants to know the height of a tree in her garden. She measures the length of the shadow of the tree and her own shadow.

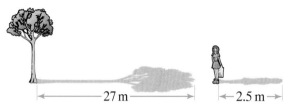

27 m ← → 2.5 m

If Sylvie is 160 cm tall, calculate the height of the tree in metres.

11 **Get Real!**

At a particular time of day the furthest points of the shadows from a tower block and a flag pole meet exactly at the point X.

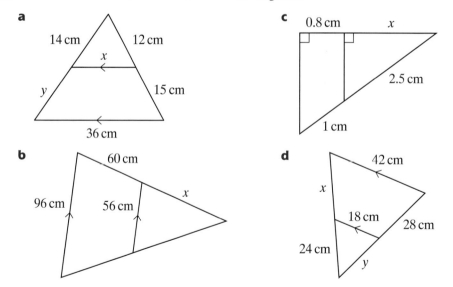

X

← 44 m → ← 61.6 m →

Joe, the architect, measures the lengths of the shadows.
He knows the building is 54 m high.
What is the height of the flagpole?

12 Find the sides marked with letters in these diagrams.

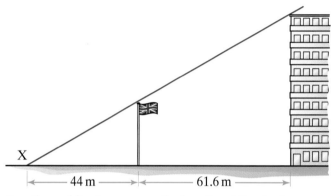

a

14 cm   12 cm
x
y 15 cm
36 cm

b

60 cm
96 cm   56 cm x

c

0.8 cm x
2.5 cm
1 cm

d

42 cm
x
18 cm   28 cm
24 cm y

Explore

◎ Produce a pattern of radiating rectangles
If the shorter side has length $2k + 1$, then the longer side has length $4k + 2$
Substitute different values of k

◎ Are your rectangles similar?

◎ Try writing your own algebraic expressions to link the sides of rectangles

◎ Which sets of rectangles are similar?

Investigate further

Explore

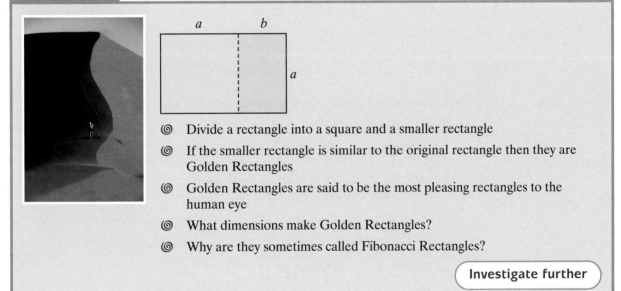

◎ Divide a rectangle into a square and a smaller rectangle

◎ If the smaller rectangle is similar to the original rectangle then they are Golden Rectangles

◎ Golden Rectangles are said to be the most pleasing rectangles to the human eye

◎ What dimensions make Golden Rectangles?

◎ Why are they sometimes called Fibonacci Rectangles?

Investigate further

Learn 2   Areas and volumes of similar shapes

Examples:

A gift shop sells small candles that are 8 cm high and cost £3.20
It sells similar candles that are 14 cm high.

a What price are the large candles if the cost of the candle is proportional
to its volume?

b It takes 144 cm^2 of plastic film to wrap a small candle.
How much does a large candle need?

a If the linear ratio of two solids is $a : b$ — This is the ratio of the corresponding lengths

then the area ratio will be $a^2 : b^2$

and the volume ratio will be $a^3 : b^3$

Linear ratio is $8 : 14 = 4 : 7$ ← Simplify the ratio

Volume ratio is $4^3 : 7^3 = 64 : 343$ ← Cube both parts for the volume ratio

Let p be the price of a large candle.
The cost ratio is $p : 3.2$

— 3.2 is the cost in £ of the small candle

Cost of large candle $\dfrac{p}{3.2} = \dfrac{343}{64}$ Volume of large candle
Cost of small candle   Volume of small candle

$$3.2 \times \frac{p}{3.2} = \frac{343}{64} \times 3.2$$

$$p = \frac{343 \times 3.2}{64}$$

$$p = £17.15$$

— Check that the price is more for the larger candle

b The amount of plastic film is proportional to the surface area of the candle.

Linear ratio is $8 : 14 = 4 : 7$

Area ratio is $4^2 : 7^2 = 16 : 49$ ← Square both parts for the area ratio

Let a be the amount of plastic film needed for
a large candle. The area ratio is $a : 144$.

Film for large candle $\dfrac{a}{144} = \dfrac{49}{16}$ Surface area of large candle
Film for small candle   Surface area of small candle

$$144 \times \frac{a}{144} = \frac{49}{16} \times 144$$

$$a = \frac{49 \times 144}{16}$$

$$a = 441 \text{ cm}^2$$

— Check that the area is more for the larger candle

Similar shapes have corresponding lengths in the same ratio.
Their areas and volumes will be in related ratios.

Apply 2

1 These pairs of solids are similar.

   i Work out the ratio of the surface area of each pair.

  ii Work out the ratios of their volumes.

Write the ratios in their simplest form, $a:b$.

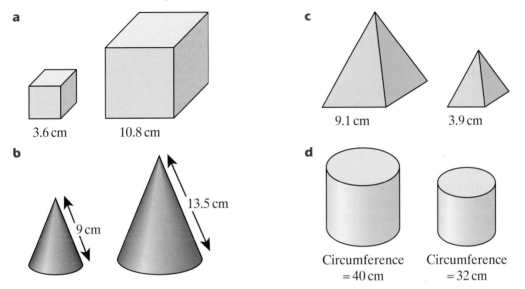

a

3.6 cm   10.8 cm

b

9 cm   13.5 cm

c

9.1 cm   3.9 cm

d

Circumference = 40 cm   Circumference = 32 cm

2 The lengths of two similar triangular prisms are in the ratio $2:7$.
The volume of the larger prism is 4459 cm^3.
What is the volume of the smaller prism?

3 The sides of two similar rhombuses are in the ratio $9:4$.
The area of the larger rhombus is 129.6 cm^2.
What is the area of the smaller rhombus?

4 Two spheres have volumes of 405 cm^3 and 120 cm^3.

  a Write the ratio of their volumes in its simplest form.

  b Find the ratio of their radii.

  c What is the ratio of their surface areas?

5 Get Real!

Two similar bottles of Quango, a fizzy drink, are 15 cm and 21 cm high.
The smaller bottle holds 500 mℓ. How much does the larger bottle hold?

6 Get Real!

The original Statue of Liberty in New York is 140 feet high and weighs
450 000 pounds.
There is a miniature version in Paris which is similar to the original and
is 35 feet high.
Assuming both statues are made of the same material, how much does
the Parisian statue weigh?

7 Get Real!

Hajra needs a new trampoline. Her old trampoline has an area of 7850 cm². There are two (similar) sizes available. One is half the width of her old one. The other is double the width of her old one.

a What is the area of the one with half the width?

b What is the area of the one with double the width?

8 Get Real!

The famous actor, Sullivan Ellis, took his gold Toscar award and melted it down to make 27 similar awards – one for each bathroom of his mansion. The miniature versions were 12 cm high. How tall was the original Toscar?

9 Get Real!

A monument has a sphere on top.

The town council wants to have the sphere, diameter 60 cm, covered in gold leaf. The gold leaf to cover a sphere of diameter 1 m costs £784. How much will the gold leaf for the smaller 60 cm sphere cost?

10 Get Real!

A supermarket sells two sizes of mustard in similar jars. The smaller size has a net weight of 250 g and the area of the lid is 12.5 cm². If the larger size has a net weight of 400 g, what is the area of the lid?

Explore

◎ A sheet of A4 paper can be cut in half to make two sheets of A5 paper

◎ Is a sheet of A4 similar to a sheet of A5?

◎ Find out about paper sizes

Investigate further

Explore

◎ A cuboid has sides x cm, y cm and z cm

◎ A second cuboid has sides $2x$ cm, $2y$ cm and $2z$ cm

◎ What is the ratio of their corresponding sides?

◎ Work out the surface area of each cuboid
What is the ratio of their surface areas?

◎ Work out the volume of each cuboid
What is the ratio of their volumes?

◎ A third cuboid has sides $3x$ cm, $3y$ cm and $3z$ cm

z cm

y cm

x cm

Investigate further

Learn 3   Proving congruence

Examples:

a Prove that triangles ABC and BCD are congruent.

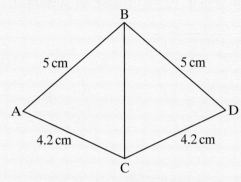

b Prove that triangles ABC and CDE are congruent.

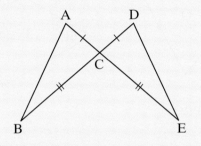

a In triangle ABC and triangle BCD:

AB = DB   (given)
AC = DC   (given)
BC = BC   (same line)

So   triangle ABC ≡ triangle DBC (SSS)

The symbol ≡ means 'is congruent to'

As with similar triangles, the order of the letters is important Triangle ABC is congruent to triangle DBC. This means that angle A = angle D, angle B = angle B and angle C = angle C, AB = DB and AC = DC

b In triangle ABC and triangle CDE:

AC = DC   (given)
BC = EC   (given)
angle ACB = angle DCE   (opposite angles)

So triangle ACB ≡ triangle DCE   (S A S)

The proof with 2 sides and 1 angle only works if the angle is between the two sides (SAS)

To prove that two triangles are congruent (have the same size and shape) you need to show one of these sets of conditions:

- both triangles have 3 corresponding sides equal (SSS)
- both triangles have 2 angles and 1 corresponding side equal (AAS)
- both triangles have 2 corresponding sides and the angle between those sides equal (SAS)
- both triangles have a right angle, hypotenuse and another side equal (RHS).

Apply 3

1 These triangles are not drawn accurately. Which triangles are congruent to triangle A?

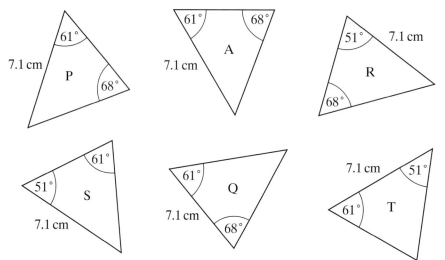

2 Triangle ABC is congruent to triangle PRQ.

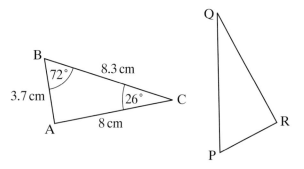

Find:

a angle Q

b RQ

c angle P.

3 Triangle BOY is congruent to triangle GRL.
Angle B = 63°, angle Y = 47°, OY = 4.6 cm and BY = 4.9 cm.
Find:

a angle L

b GL

c angle R.

4 Amy says that since AC = QP, AB = QR and angle A = angle Q,
triangle ABC is congruent to triangle PQR (SAS).
Laura says that Amy has made a mistake.
Who is right?
Give a reason for your answer.

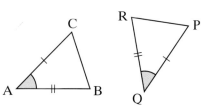

5 Get Real!
Lifeguard X and Lifeguard Y each spot a child in the sea from two different
lifeguard stations on the beach. The diagram shows their relative positions
and the position of a flag, F, which is equidistant from both lifeguards.
C is the position of the child and B is a marker buoy out to sea.

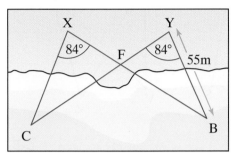

By first proving that triangles CXF and BYF are congruent,
find the distance from Lifeguard X to the child.

6 Look at the following pairs of triangles, which are not drawn accurately.
If they are congruent give the reason and name the other pairs of equal
sides and angles.

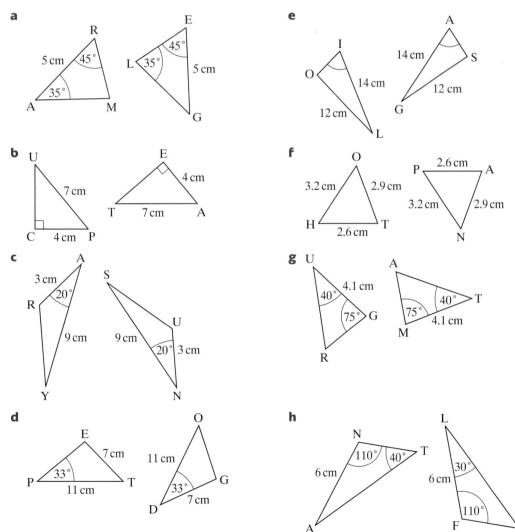

7 Prove that each of these diagrams contains a pair of congruent triangles.
Some equal sides and angles are marked.

a

b

c

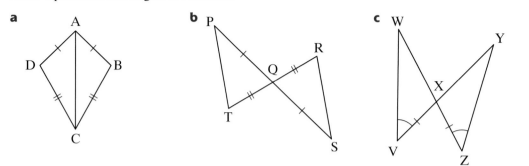

8 The diagram shows the construction of an angle bisector.
Prove the construction theorem by following these steps.

a Prove triangle AXZ ≡ triangle AYZ.

b Hence show that angle XAZ = angle YAZ.

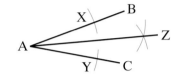

You need to understand and learn
these two proofs for the exam

9 The diagram shows the construction of a perpendicular bisector.
Prove the construction theorem by following these steps.

a Prove that triangle ACD ≡ triangle BCD.

b Hence show that angle ACD = angle BCD.

c Now prove that triangle ACX ≡ triangle BCX.

d Hence show that AX = BX and angle CXA = angle CXB = 90°

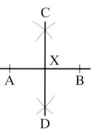

10 In the diagram, ABC, ADE and CEF are all equilateral triangles of
different sizes.

a Prove that triangle ABD ≡ triangle ACE.

b Can you find a third triangle congruent to ABD and ACE?

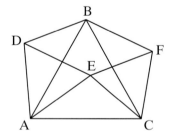

11 Get Real!

The owner of Nice Ices is putting this symbol on all of her
ice cream parlours.

ABCDE is a regular pentagon.

Prove that triangle BDA ≡ triangle EFA.

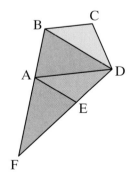

Explore

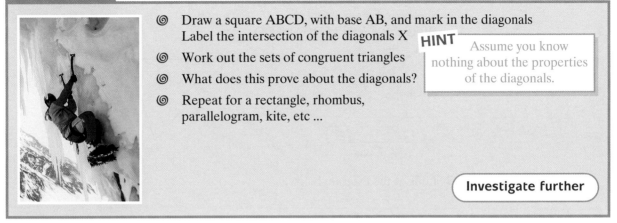

- ⊚ Draw a square ABCD, with base AB, and mark in the diagonals
  Label the intersection of the diagonals X
- ⊚ Work out the sets of congruent triangles
- ⊚ What does this prove about the diagonals?
- ⊚ Repeat for a rectangle, rhombus,
  parallelogram, kite, etc ...

HINT Assume you know nothing about the properties of the diagonals.

Investigate further

Explore

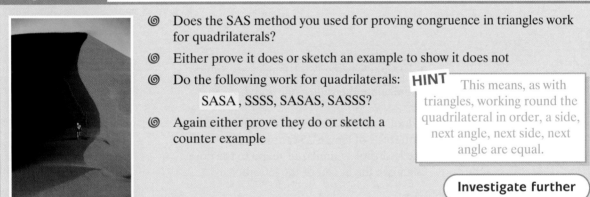

- ⊚ Does the SAS method you used for proving congruence in triangles work for quadrilaterals?
- ⊚ Either prove it does or sketch an example to show it does not
- ⊚ Do the following work for quadrilaterals:
  SASA , SSSS, SASAS, SASSS?
- ⊚ Again either prove they do or sketch a counter example

HINT This means, as with triangles, working round the quadrilateral in order, a side, next angle, next side, next angle are equal.

Investigate further

Similarity and congruence

ASSESS

The following exercise tests your understanding of this chapter, with the questions appearing in order of increasing difficulty.

1 Triangle WIL is congruent to triangle YNG. Write down the sizes of all the angles and sides in triangle YNG.

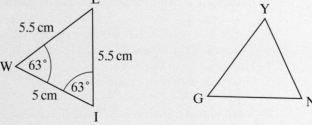

2 Triangle CAT is congruent to triangle DOG. Write down the sizes of all the angles of triangle DOG.
Write the lengths of DG and DO.

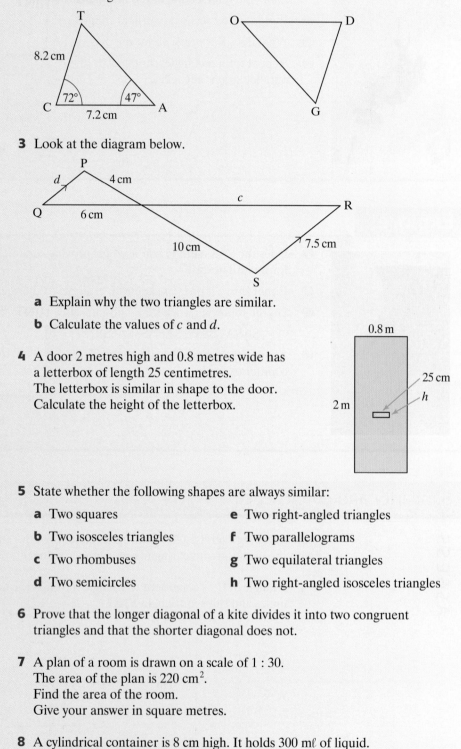

3 Look at the diagram below.

a Explain why the two triangles are similar.

b Calculate the values of c and d.

4 A door 2 metres high and 0.8 metres wide has a letterbox of length 25 centimetres.
The letterbox is similar in shape to the door.
Calculate the height of the letterbox.

5 State whether the following shapes are always similar:

a Two squares

b Two isosceles triangles

c Two rhombuses

d Two semicircles

e Two right-angled triangles

f Two parallelograms

g Two equilateral triangles

h Two right-angled isosceles triangles

6 Prove that the longer diagonal of a kite divides it into two congruent triangles and that the shorter diagonal does not.

7 A plan of a room is drawn on a scale of 1 : 30.
The area of the plan is 220 cm^2.
Find the area of the room.
Give your answer in square metres.

8 A cylindrical container is 8 cm high. It holds 300 mℓ of liquid.
A similar container is 20 cm high.
How much liquid will the larger container hold?

C ▷ **Examiners would normally expect students who get a C grade to be able to:**

Use Pythagoras' theorem to find the hypotenuse of a right-angled triangle

Use Pythagoras' theorem to find any side of a right-angled triangle

Use Pythagoras' theorem to find the height of an isosceles triangle

Use Pythagoras' theorem in practical problems

B ▷ **Examiners would normally expect students who get a B grade also to be able to:**

Find the distance between two points from their coordinates

A ▷ **Examiners would normally expect students who get an A grade also to be able to:**

Use Pythagoras' theorem in 3-D problems

What you should already know ...

■ Squares of integers up to 15 and the corresponding square roots

■ Round numbers with decimals to the nearest integer

■ The properties of quadrilaterals and their diagonals

■ Calculate areas and volumes

■ Circle properties

■ Use and simplify surds

■ Plot 3-D coordinates

Equilateral triangle – a triangle with 3 equal sides and 3 equal angles – each angle is 60°

Isosceles triangle – a triangle with 2 equal sides and 2 equal angles; the equal angles are called **base angles**

Right-angled triangle – a triangle with one angle of 90°

Hypotenuse – the longest side of a right-angled triangle, opposite the right angle

Hypotenuse

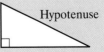

Surd – a number containing an irrational root, for example, $\sqrt{2}$ or $3 + 2\sqrt{7}$

Pythagoras' theorem – in a right-angled triangle, the square of the length of the hypotenuse is equal to the sum of the squares of the lengths of the other two sides

$$c^2 = a^2 + b^2$$

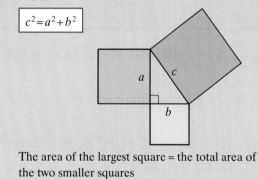

The area of the largest square = the total area of the two smaller squares

Converse of Pythagoras' theorem – in any triangle, if $c^2 = a^2 + b^2$ then the triangle has a right angle opposite c; for example, if $c = 17$ cm, $a = 8$ cm, $b = 15$ cm, then $c^2 = a^2 + b^2$ so this is a right angle

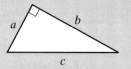

Pythagorean triple – a set of three integers a, b, c that satisfies $c^2 = a^2 + b^2$; for example, 3, 4, 5 $(5^2 = 3^2 + 4^2)$, 5, 12, 13 $(13^2 = 5^2 + 12^2)$, 6, 8, 10 $(10^2 = 6^2 + 8^2)$ and 15, 36, 39 $(39^2 = 15^2 + 36^2)$

Learn 1   Pythagoras' theorem in 2-D

Examples: **a** Find AC, leaving your answer as a square root.

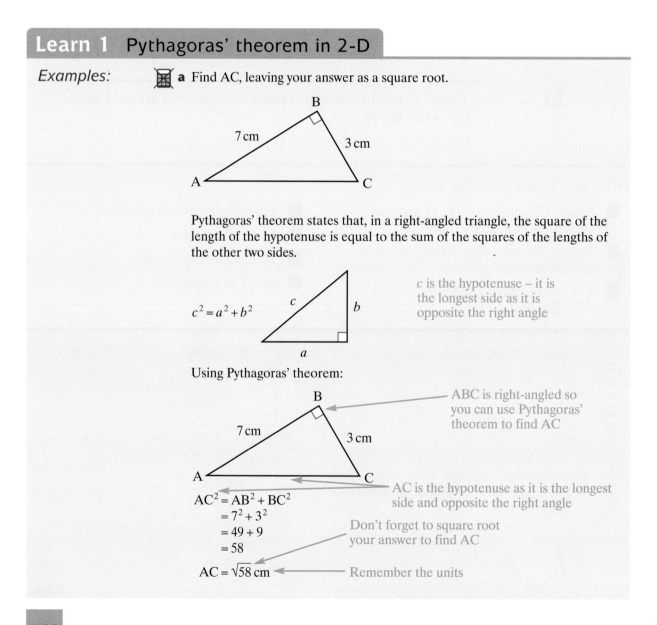

Pythagoras' theorem states that, in a right-angled triangle, the square of the length of the hypotenuse is equal to the sum of the squares of the lengths of the other two sides.

$$c^2 = a^2 + b^2$$

c is the hypotenuse – it is the longest side as it is opposite the right angle

Using Pythagoras' theorem:

ABC is right-angled so you can use Pythagoras' theorem to find AC

$$AC^2 = AB^2 + BC^2$$
$$= 7^2 + 3^2$$
$$= 49 + 9$$
$$= 58$$
$$AC = \sqrt{58} \text{ cm}$$

AC is the hypotenuse as it is the longest side and opposite the right angle

Don't forget to square root your answer to find AC

Remember the units

b Find PQ, giving your answer to an appropriate degree of accuracy.

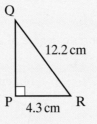

Using Pythagoras' theorem:

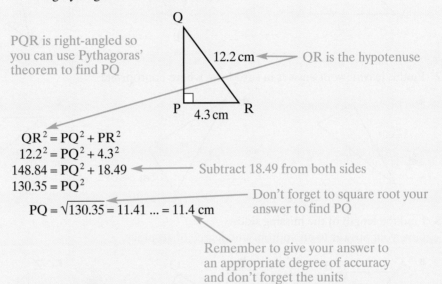

PQR is right-angled so you can use Pythagoras' theorem to find PQ

QR is the hypotenuse

$QR^2 = PQ^2 + PR^2$
$12.2^2 = PQ^2 + 4.3^2$
$148.84 = PQ^2 + 18.49$ ← Subtract 18.49 from both sides
$130.35 = PQ^2$

$PQ = \sqrt{130.35} = 11.41\ldots = 11.4\text{ cm}$

Don't forget to square root your answer to find PQ

Remember to give your answer to an appropriate degree of accuracy and don't forget the units

c Find the length of the line AB, where A is the point $(-1, 1)$ and B is the point $(4, 5)$.

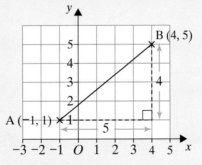

Using Pythagoras' theorem:

$AB^2 = 5^2 + 4^2$
$AB^2 = 25 + 16$
$AB^2 = 41$
$AB = \sqrt{41}$
$AB = 6.4\text{ units}$

Apply 1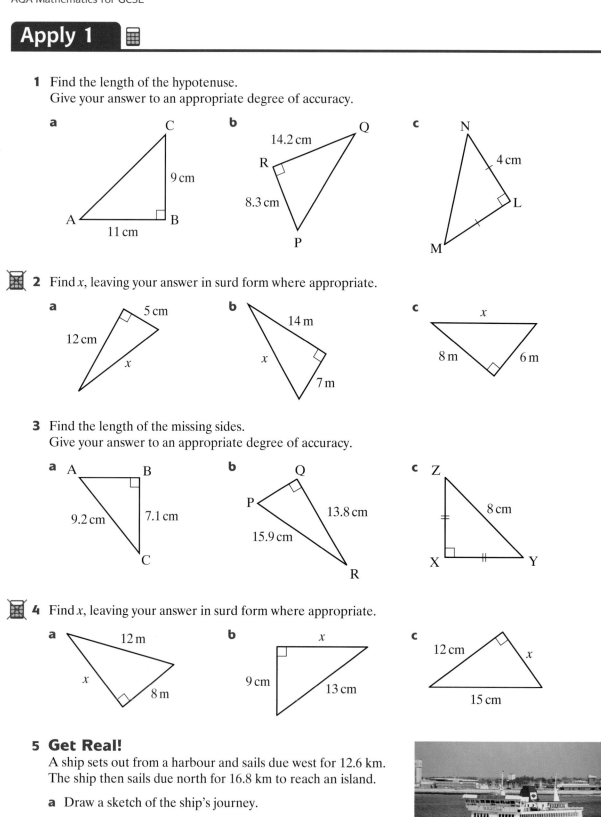

1 Find the length of the hypotenuse.
Give your answer to an appropriate degree of accuracy.

a

C
9 cm
A
11 cm
B

b

14.2 cm
R
Q
8.3 cm
P

c

N
4 cm
L
M

2 Find x, leaving your answer in surd form where appropriate.

a

5 cm
12 cm
x

b

14 m
x
7 m

c

x
8 m
6 m

3 Find the length of the missing sides.
Give your answer to an appropriate degree of accuracy.

a

A
B
9.2 cm
7.1 cm
C

b

Q
P
13.8 cm
15.9 cm
R

c

Z
8 cm
X
Y

4 Find x, leaving your answer in surd form where appropriate.

a

12 m
x
8 m

b

x
9 cm
13 cm

c

12 cm
x
15 cm

5 Get Real!
A ship sets out from a harbour and sails due west for 12.6 km.
The ship then sails due north for 16.8 km to reach an island.

a Draw a sketch of the ship's journey.

b Calculate the shortest distance from the harbour to the island.

6 Albert is trying to find x.

This is his working:

$x^2 = 6^2 + 4^2$
$x^2 = 36 + 16$
$x^2 = 52$
$x = \sqrt{52}$ cm

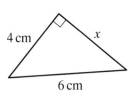

4 cm x

6 cm

Is Albert correct?
Give a reason for your answer.

7 O is the centre of a circle of radius 9 cm.
X is a point on the circumference of the circle and OXY is a straight line.
The length of the tangent TY is 12 cm.

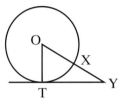

O

X

T   Y

What is the length of XY?

8 Get Real!

ABCD is the lid of a rectangular pencil tin.

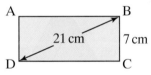

A _____ B

21 cm   7 cm

D _____ C

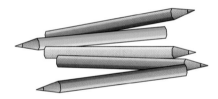

What is the area of the lid?
Give your answer to three significant figures.

9 ABC is an isosceles triangle with AB = AC and vertical height 4 cm.
Mr Armitage asks his class to find AC.

Gareth says that you can't use Pythagoras' theorem because ABC is an isosceles triangle and is not right-angled.

Ruth disagrees. She says that there is a right angle and if you use Pythagoras' theorem you get $AC^2 = 4^2 + 3^2 = 25$, so AC = 5 cm.

Simon says that Ruth has also made a mistake and that $AC = 4^2 + 1.5^2 = 18.25$ cm.

In fact, all three are wrong!

a Explain why there is a right-angled triangle.

b What did Ruth and Simon do wrong?

c Use Pythagoras' theorem correctly to find AC.

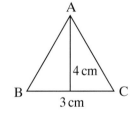

A

4 cm

B   C

3 cm

10 Get Real!

Annette is designing a sun room with a sloping glass roof.
The diagram shows a side elevation of the room with her planned measurements.

Calculate the length of the sloping roof.
Give your answer in metres and centimetres.

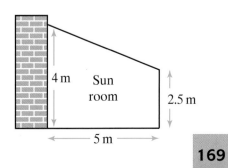

4 m   Sun room

2.5 m

5 m

169

11 Get Real!

The safety instructions with a 6.1 metre ladder say that the foot of the ladder must be a maximum distance of 1.7 metres and a minimum distance of 1.5 metres from the wall.

To the nearest centimetre, what is the maximum vertical height that the ladder can safely reach?

12 A right-angled triangle has a hypotenuse of length 25 cm.
The lengths of the other two sides are whole numbers of centimetres.
Can you find two different possible triangles?

13 Sketch a diagram and use Pythagoras' theorem to find the distance between each of the following pairs of points. Give each length to two significant figures.

   a A(2, 4), B(7, 6)

   b C(1, 5), D(4, 2)

   c P(−2, 1), Q(3, 8)

   d X(−3, 6), Y(1, −2)

14 Draw accurately a line of length:

   a $\sqrt{13}$ cm   **b** $\sqrt{17}$ cm

15 Get Real!

Isobel is working out the volume of a chocolate mint that is in the shape of a triangular prism.

She knows that the volume of a prism = area of cross-section × length
She calculates the volume as 3 cm × 3 cm × 0.5 cm = 4.5 cm^3
Is Isobel correct?
Give a reason for your answer.

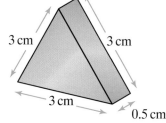

16 Haseeb works out the perimeter of the quadrilateral ABCD.
He uses Pythagoras' theorem in triangle ABC to calculate AC.
Then he uses Pythagoras' theorem in triangle ADC to find DC.
Haseeb says that DC is 2 cm and the perimeter is 24 cm.
Is Haseeb correct?
Give a reason for your answer.

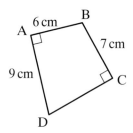

17 Get Real!

A boat has two masts that are 3.7 m apart.
The smaller mast is 5.5 m high and is 1.8 m from the front of the boat.
The taller mast is 8.5 m high and is 2.9 m from the back of the boat.
Kasim wants to attach a string of fairy lights from the back of the boat to the front of the boat, going over the tops of the masts.

To the nearest metre, what is the shortest string of lights that he will need?

18 A rhombus has a perimeter of 40 cm.
Its shorter diagonal is 12 cm.

   a Find the length of the longer diagonal of the rhombus.

   b What is the area of the rhombus?

19 Get Real!

Catriona has designed a jade pendant for a necklace.
It is a kite, with the measurements shown in the diagram.
The width of the pendant is 4 cm.

What is the height, x, of the pendant to the nearest millimetre?

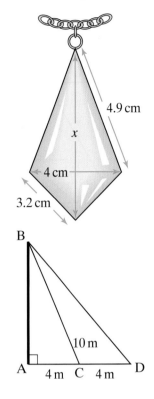

20 Get Real!

A transmitter, AB, is supported by two guy ropes, BC and BD.
Angle BAC = 90°
The shorter guy rope, BC, is 10 m.
AC = CD = 4 m

Show that the length of the longer guy rope, BD, is $\sqrt{148}$ m.

21

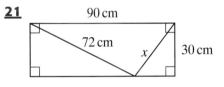

Calculate x, giving your answer to the nearest millimetre.

22 ABCD is an isosceles trapezium, with base DC = 43 cm

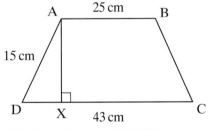

Not drawn
accurately

Calculate the area of ABCD.

23 Get Real!

A farmer is fencing off the central area of his field as shown in the diagram.
The surrounding field is a square of side 120 metres.

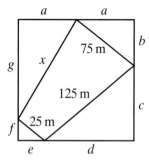

a State the value of a.

b Find the values of b, c, d, e, f and g.

c Calculate the length, x, of the fourth side of the enclosure.

Give your answers to the nearest metre.

24 Get Real!

The diagram shows the side view of the rain cover on Marianna's toy pram.
Find the length of a, leaving your answer as a square root.

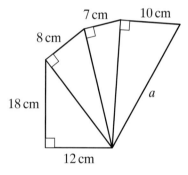

25 Get Real!

A graphic designer has drawn the start of a phone number.
The number eight outline is formed from two overlapping yellow circles.
The top circle has centre X and radius 13 cm.
The bottom circle has centre Y and radius 15 cm.
AB, the width across the narrowest part of the eight, is 24 cm.

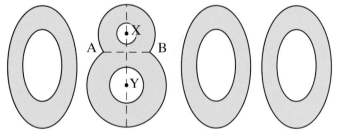

Calculate XY and hence find the height of the number eight.

26 Using the area of the triangle to help you, calculate x.

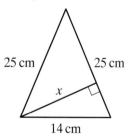

27 Adam is buying a plasma screen television for the wall of
his penthouse apartment.
The screen size, measured along the diagonal, is 42 inches.
The width to height ratio of the screen is 16 : 9.
Find the width and the height of Adam's television, giving
your answers to three significant figures.

Explore

◎ Construct these triangles accurately using a ruler and a pair of compasses
   i   AB = 8 cm, BC = 6 cm, AC = 10 cm
   ii AB = 7.4 cm, BC = 4.6 cm, AC = 8.7 cm
   iii AB = 10.2 cm, BC = 5.1 cm, AC = 8.8 cm
   iv AB = 6.1 cm, AC = BC = 4.3 cm

◎ What kind of triangles are they? Use a protractor to check

◎ How could you have shown this would be true without constructing the
triangles?

◎ Using Pythagoras' theorem 'in reverse' is called the **converse** of Pythagoras'
theorem

◎ Work out which of the following lengths would form right-angled triangles
   i   3 cm, 4 cm, 5 cm
   ii 6 cm, 7 cm, 8 cm
   iii 5 cm, 12 cm, 13 cm
   iv 2 cm, $\sqrt{8}$ cm, $\sqrt{12}$ cm
   v $3\sqrt{2}$ cm, 6 cm, $6\sqrt{2}$ cm

◎ Take some measurements from objects in the room, for example, the door,
the window, a table, a chair, a book, etc.

◎ Use the lengths of the diagonals and the converse of Pythagoras' theorem
to check that they have a vertex that is a right angle

> **Investigate further**

Explore

◎ In Egypt the farmers who farmed the fields on the banks of the Nile used
periodically to have the boundaries of their fields washed away by floods; to
mark out their fields again they used a rope divided into 12 equal sections
by knots – how was this used?

◎ Also in Ancient Egypt, the surveyors were known as rope-stretchers
because they used the above technique to mark out the Pyramids

◎ Can you discover any other historical uses of the converse of Pythagoras'
theorem, or any modern day uses?

> **Investigate further**

Explore

◎ Sets of positive integers that satisfy Pythagoras' theorem are called Pythagorean triples

◎ Euclid (another famous mathematician) found that if you take two positive integers x and y, you can generate a Pythagorean triple of the form $2xy, x^2 - y^2, x^2 + y^2$

◎ Which Pythagorean triple is generated when $x = 7$ and $y = 1$?

◎ Experiment with Euclid's formula by generating more triples (you could use a spreadsheet)

◎ Can you show that Euclid's formula will always generate a triple? Are there any conditions on your choice of x and y?

◎ The triple generated by $x = 7$ and $y = 1$ is called a **multiple**. The **primitive** triple is 7, 24, 25. What do you think the difference is between primitive triples and multiples?

Investigate further

Explore

◎ Pythagoras' theorem states that, in a right-angled triangle, $c^2 = a^2 + b^2$

◎ What does that tell you about the areas of the coloured squares, A_1, A_2 and A_3?

◎ This diagram is a famous representation of Pythagoras' theorem

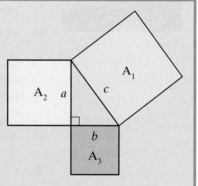

◎ If the squares were changed into semicircles, would Pythagoras' theorem still work for the areas?
Can you show this algebraically?

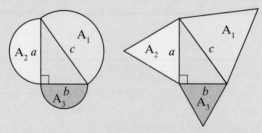

◎ What happens if the squares are replaced by equilateral triangles?

◎ Try replacing the squares with any set of similar shapes

Investigate further

Learn 2   Pythagoras' theorem in 3-D

Examples:

a What is the length of the longest straight rod that will fit into the box shown in the diagram?

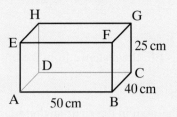

The longest rod that will fit will lie along the diagonal AG.
AG is the hypotenuse of triangle ACG.

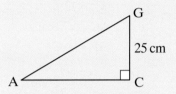

First we need to find AC.
Using Pythagoras' theorem in the right-angled triangle ABC:

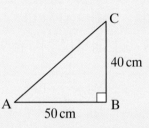

$$AC^2 = AB^2 + BC^2$$
$$AC^2 = 50^2 + 40^2$$
$$AC^2 = 4100$$

You need this in the next calculation, so leave your answer in surd form

$$AC = \sqrt{4100} \text{ cm}$$

Using Pythagoras' theorem in the right-angled triangle ACG:

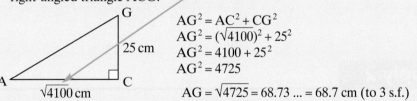

$$AG^2 = AC^2 + CG^2$$
$$AG^2 = (\sqrt{4100})^2 + 25^2$$
$$AG^2 = 4100 + 25^2$$
$$AG^2 = 4725$$
$$AG = \sqrt{4725} = 68.73 \ldots = 68.7 \text{ cm (to 3 s.f.)}$$

b The square-based pyramid shown in the diagram is made with four isosceles triangles each with sides 8 cm, 8 cm and 6 cm.

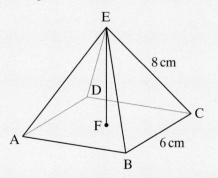

What is the vertical height of the pyramid?
Leave your answer in surd form.

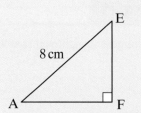

The vertical height is EF, the distance from the top of the pyramid to the centre of the base.

So first find AF, where $AF = \frac{1}{2}AC$

Using Pythagoras' theorem in the right-angled triangle ABC:

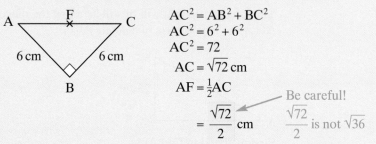

$AC^2 = AB^2 + BC^2$
$AC^2 = 6^2 + 6^2$
$AC^2 = 72$
$AC = \sqrt{72}$ cm
$AF = \frac{1}{2}AC$

$= \frac{\sqrt{72}}{2}$ cm

Be careful!
$\frac{\sqrt{72}}{2}$ is not $\sqrt{36}$

Using Pythagoras' theorem in the right-angled triangle AFE:

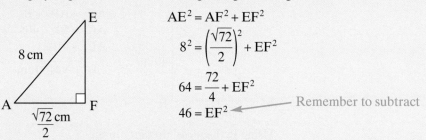

$AE^2 = AF^2 + EF^2$

$8^2 = \left(\frac{\sqrt{72}}{2}\right)^2 + EF^2$

$64 = \frac{72}{4} + EF^2$

$46 = EF^2$

Remember to subtract

So the vertical height of the pyramid, $EF = \sqrt{46}$ cm

Apply 2

1 The diagram shows a cuboid in which AE = 5 cm, AB = 12 cm and BC = 8 cm.

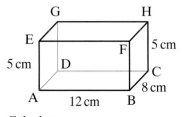

Calculate:

a AC

b CE

c XE, where X is the midpoint of AC.

2 Find the length of the longest straight rod that could fit in each of these boxes.
Leave your answers in surd form where appropriate.

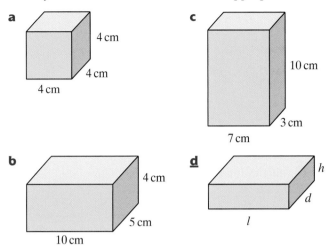

a

4 cm
4 cm
4 cm

c

10 cm
3 cm
7 cm

b

4 cm
5 cm
10 cm

d

h
d
l

3 Get Real!

Romeo asks the florist to send Juliet a single red rose in a clear box
with a ribbon around it.

The florist puts the ribbon diagonally up the front of the box, straight
across the top edge, diagonally down the back and then along the
bottom edge to meet up.

How much ribbon does the florist use?

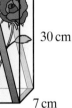

30 cm
7 cm
7 cm

4 OABCDEFG is a cube of side 5 cm.
O is the origin.

a What are the coordinates of the point B?

b Calculate the distance OB.

c Calculate the distance OF.

d Which other dimensions of the cube are
equal to OF?

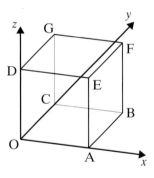

z G y
F
D
E
C
B
O
A x

5 ABCDEFGH is a cube of side 20 cm.
X is the midpoint of CG.
Mr Cowell tells his class to calculate AX.
Sharon uses Pythagoras' theorem in triangles ABC and ACX.
Louis uses Pythagoras' theorem in triangles BCX and ABX.

a Whose method is correct? Give reasons for your answer.

b What is the length of AX?

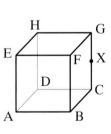

H G
E F X
D C
A B

6 Get Real!

The diagram shows a skateboard ramp in a park.
The ramp is a triangular prism with dimensions as shown.
Maya uses her skateboard to travel down the slope in a straight line
from Y to A.
How far does Maya travel?

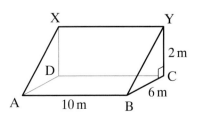

X Y
2 m
D
C
A 10 m B 6 m

177

7 Albert finds the surface area of a triangular prism as follows:

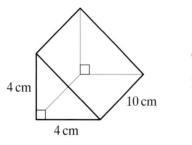

Triangle end area $= \frac{1}{2} \times 4 \times 4 = 8 \text{ cm}^2$

Rectangle base area $= 10 \times 4 = 40 \text{ cm}^2$

4 cm

10 cm

4 cm

Albert notices that the triangular prism is made of 2 triangles and 3 rectangles so:

Surface area = 2 ends + 3 sides = $(2 \times 8) + (3 \times 40) = 16 + 120 = 136 \text{ cm}^2$

Is Albert correct?

Give a reason for your answer.

8 Get Real!

A child's magnetic fishing game comes in a cuboid box.

The fishing rod is a 13 cm stick that just fits in the box.

Given that the dimensions of the box are all integers, find them.

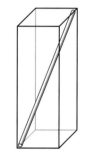

9 Get Real!

The manufacturer of a chocolate bar wants to make a special chocolate orange edition with a diagonal stripe on one side of the box.

The box is in the shape of a triangular prism.

Using the dimensions in the diagram, calculate the length of the diagonal stripe.

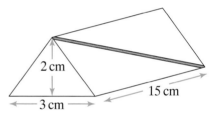

2 cm

3 cm

15 cm

10 ABCDE is a rectangular-based pyramid.

EF = 8 cm

Find the slant height EC.

Give your answer in surd form.

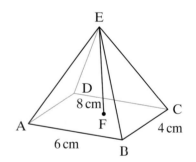

E

D

8 cm

C

A

F

4 cm

6 cm

B

11 PQRST is a rectangular-based pyramid.

RT = 10 cm

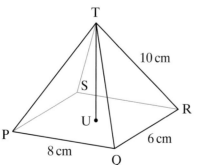

T

10 cm

S

R

U

P

6 cm

8 cm

Q

Find the perpendicular height TU.

Give your answer in surd form in its lowest terms.

12 Get Real!

Annette's sun room is finished.
It is a prism with a right-angled trapezium for the
cross-section.
Annette measures the diagonals to check that the
builders have done a good job.
If the room has been built accurately, what should
the diagonals AG and DF measure?

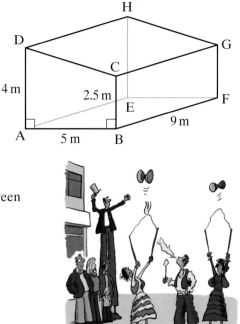

13 Get Real!

A diabolo is a spinning top made of two equal open cones.
It is spun and thrown by jugglers using a piece of cord between
two sticks.

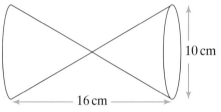
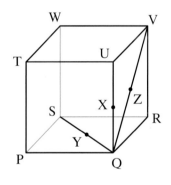

Find the total outer curved surface area of the diabolo in the diagram.
Use the fact that the surface area of a cone is $\pi r l$ where r is the radius
and l is the slant height.

14 PQRSTUVW is a wire cube of side 10 cm.
X, Y and Z are the midpoints of QU, QS and QV.
Which of the following lengths are correct?

a $QY = 5\sqrt{2}$ cm   **d** $XY = 5\sqrt{3}$ cm

b $QW = \sqrt{210}$ cm   **e** $YZ = 10$ cm

c $QZ = 5\sqrt{5}$ cm   **f** $SZ = 5\sqrt{10}$ cm

Find the correct lengths for those that are wrong.

15 Get Real!

The diagram shows a room that is in the shape of a cuboid.
A spider is in the middle of the edge of the ceiling at S.
A fly is in the bottom corner of the room at F.

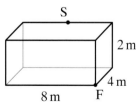

a The spider crawls by the most direct route from S to F across the
ceiling and down the wall, to try to catch the fly.
What distance does the spider travel?
You may wish to draw a net of the room to help you.

b At the same time the fly avoids the spider by flying directly from F to S.
What distance does the fly travel?

16 Get Real!

The glass pyramid in the Louvre museum is a square-based pyramid of base edge 38 metres.
The vertex of the pyramid is directly over the midpoint of the base.
The volume of the pyramid is approximately 10 400 cubic metres.

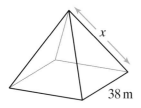

Volume of a pyramid $= \frac{1}{3} \times$ base area $\times$ height

Find the length of the slant edge of the pyramid, x.

17 ABCDEFGH is a cube of side 4 cm.
X, Y and Z are the midpoints of AE, BC and GH.

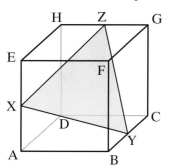

Calculate the area of the triangle XYZ.

Explore

◎ In Learn **1** you found the distance between two points from their coordinates

◎ Can you find a general rule for the distance between two points?

◎ Now think in three dimensions. How would you find the distance between the two points in the diagram?

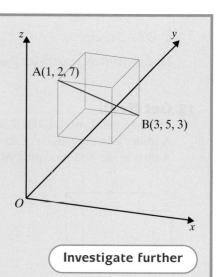

Investigate further

Pythagoras' theorem

The following exercise tests your understanding of this chapter, with the questions appearing in order of increasing difficulty.

1 Find the marked lengths in these diagrams.

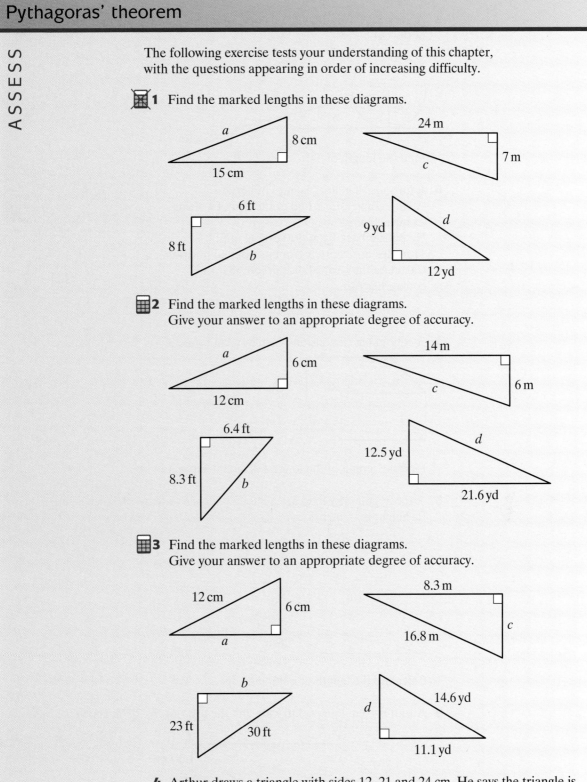

2 Find the marked lengths in these diagrams.
Give your answer to an appropriate degree of accuracy.

3 Find the marked lengths in these diagrams.
Give your answer to an appropriate degree of accuracy.

4 Arthur draws a triangle with sides 12, 21 and 24 cm. He says the triangle is right-angled. Is he right? Give a reason for your answer.

5 A tangent, PT, is drawn to a circle of radius 3 m. P is 7.5 m from the centre of the circle.

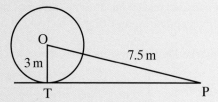

Find the length of TP.

6 John and Allan are playing conkers.
Allan's conker, C, is tied to the end of a string 25 inches long.
He pulls it back from the vertical until it is 11 inches horizontally from its original position.
Calculate the vertical distance, h, that the conker has risen.

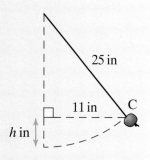

7 A cuboid has dimensions 4 cm by 5 cm by 6 cm.

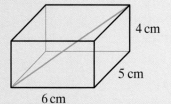

Find the length of the longest diagonal across the box.

8 A rectangular pyramid has a base 7.6 inches wide by 10.3 inches long. Its height is 15 inches.

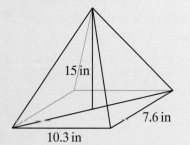

Calculate the length of a sloping edge.

9 A roof is 25 m long, 4.5 m wide and the top ridge is 2.5 m above the base.

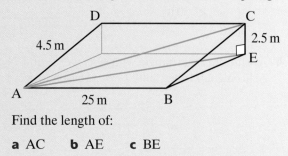

Find the length of:

a AC **b** AE **c** BE

13 Quadratic graphs

OBJECTIVES

D **Examiners would normally expect students who get a D grade to be able to:**

Draw graphs of simple quadratic functions such as $y = 3x^2$ and $y = x^2 + 4$

C **Examiners would normally expect students who get a C grade also to be able to:**

Draw graphs of harder quadratic functions such as $y = x^2 - 2x + 1$

Find the points of intersection of quadratic graphs with lines

Use graphs to find the approximate solutions of quadratic equations

A **Examiners would normally expect students who get an A grade also to be able to:**

Use the points of intersection of a quadratic graph such as $y = x^2 - 2x - 4$ with lines such as $y = 2x + 1$ to solve equations like $x^2 - 2x - 4 = 2x + 1$ and simplify this to $x^2 - 4x - 5 = 0$

What you should already know ...

- Add and subtract fractions
- Substitute positive and negative values of x into expressions including squared terms
- Plot graphs from coordinates

VOCABULARY

Variable – a symbol representing a quantity that can take different values such as x, y or z

Quadratic expression – an expression containing terms where the highest power of the variable is 2

Quadratic expressions	Non-quadratic expressions
x^2	x
$x^2 + 2$	$2x$
$3x^2 + 2$	$\frac{1}{x}$
$4 + 4y^2$	$3x^2 + 5x^3$
$(x + 1)(x + 2)$	$x(x + 1)(x + 2)$

Quadratic function – functions like $y = 3x^2$, $y = 9 - x^2$ and $y = 5x^2 + 2x - 4$ are quadratic functions; they include an x^2 term and may also include x terms and constants

The graphs of quadratic functions are always ∪-shaped or ∩-shaped

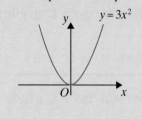

$y = 3x^2$

$y = ax^2 + bx + c$ is ∪-shaped when a is positive and ∩-shaped when a is negative

c is the intercept on the y-axis

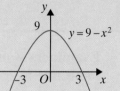

$y = 9 - x^2$

Note that other letters could be used as the variable instead of x (for example, $6t^2 - 3t - 5$ is also a quadratic expression and $h = 30t - 2t^2$ is a quadratic function)

Learn 1   Graphs of simple quadratic functions

Examples:

a Draw the graph of $y = 3x^2$ for values of x from −4 to 4.

b Use the graph to find   **i** the value of y when $x = -2.4$
                                   ii the values of x when $y = 35$

a The table below gives the values of y for all integer values of x from −4 to 4.

x	−4	−3	−2	−1	0	1	2	3	4
y	48	27	12	3	0	3	12	27	48

Quadratic functions give curves – you need more points than for linear functions to get the right shape

$3x^2$ means $3 \times x^2$ or $3 \times x \times x$ so when $x = -4$, $3x^2 = 3 \times -4 \times -4 = 48$

The y-values are often symmetrical like this

Check that you can work out the values in the table, with and without your calculator

Plotting these values gives the graph:

Graph of $y = 3x^2$

The points are joined with a smooth curve

−3.4   −2.4
Lowest (minimum) point

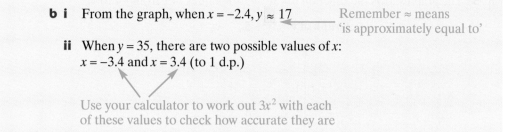

b i From the graph, when $x = -2.4$, $y \approx 17$ ← Remember $\approx$ means 'is approximately equal to'

    ii When $y = 35$, there are two possible values of x:
$x = -3.4$ and $x = 3.4$ (to 1 d.p.)

Use your calculator to work out $3x^2$ with each of these values to check how accurate they are

Apply 1

1 a Draw the graph of $y = x^2$ for values of x from -4 to 4.

    b Use your graph to estimate the value of:
      i 2.5^2 **ii** $(-1.8)^2$

    c Use your graph to estimate the square roots of:
      i 13 **ii** 5

2 a Copy and complete this table for $y = x^2 - 2$

x	-3	-2	-1	0	1	2	3
y	7		-1			2	

    b Draw the graph of $y = x^2 - 2$ for values of x from -3 to 3.

    c Use your graph to find the value of y when:
      i $x = 2.4$ **ii** $x = -1.6$

    d Use your graph to find the values of x when:
      i $y = 6$ **ii** $y = -0.5$

3 a Draw the graph of $y = -5x^2$ for values of x from -3 to 3.

    b Compare your graph with that of $y = x^2$
    What are the similarities and differences?

4 a Copy and complete the table below for $y = 2x^2 + 3$

x	-4	-3	-2	-1	0	1	2	3	4
$2x^2$	32	18	8		0			18	
$y = 2x^2 + 3$	35		11		3			21	

    b Draw the graph of $y = 2x^2 + 3$ for values of x from -4 to 4.

    c Give the y-coordinate of the point on the curve with an x-coordinate of:
      i 2.5 **ii** -1.5

    d Give the x-coordinates of the points on the curve with y-coordinates of:
      i 10 **ii** 28

5 The graph shows the function $y = 3x^2 - 5$ for values of x from -4 to 4.

Graph of $y = 3x^2 - 5$

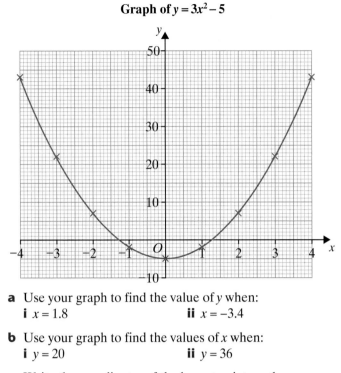

a Use your graph to find the value of y when:
   i $x = 1.8$ **ii** $x = -3.4$

b Use your graph to find the values of x when:
   i $y = 20$ **ii** $y = 36$

c Write the coordinates of the lowest point on the curve.

6 a Draw the graph of $y = 10 - x^2$ for values of x from -4 to 4.

   b i Give the x-coordinates of the points where the curve crosses
      the x-axis.

   ii Explain why the answers to part **i** are the square roots of 10.

7 a Copy and complete the table below, then use it to draw the graph of
      $y = (x + 2)(3 - x)$

x	−3	−2	−1	0	0.5	1	2	3	4
x + 2	−1		1		2.5				6
3 − x	6		4		2.5				−1
y = (x + 2)(3 − x)	−6		4		6.25				−6

b Write down the coordinates of the points where the curve crosses
   the x-axis.

8 a Draw the graph of $y = x(x - 4)$ for values of x from -1 to 5.

   b Write the coordinates of the points where the curve crosses the
      line $y = 0$.

9 Get Real!

The area of the glass in a circular window is given by $A = \pi r^2$, where r is the radius in metres and A is the area in square metres.

a Copy and complete this table, giving values of A to 2 decimal places.

r (m)	0	0.1	0.2	0.3	0.4	0.5	0.6	0.7	0.8	0.9	1.0
$A = \pi r^2$ (m²)			0.13			0.79			2.01		3.14

b Draw a graph of A against r using 2 cm to represent 0.2 on the r-axis and 0.5 on the A-axis.

c Use your graph to estimate the area of the window when the radius is:
  i 0.28 m   **ii** 0.75 m

d Use your graph to estimate the radius of the window when the area is:
  i 2.8 m²   **ii** 1.45 m²

10 The graph shows the points that Paul has plotted for his graph of $y = x^2$

Graph of $y = x^2$

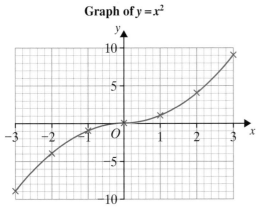

a Is this the shape you would expect?
  Give a reason for your answer.

b Complete the table of values for Paul's graph.

x	−3	−2	−1	0	1	2	3
y							

c What mistake has Paul made in calculating the values?

11 The graphs of four quadratic functions are shown in the sketch.

The functions are

$y = 6x^2$ $y = -6x^2$ $y = x^2 + 6$ $y = x^2 - 6$

Choose the function that represents each curve.

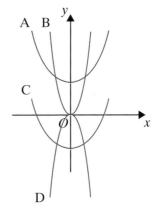

12 a On separate axes draw the graphs of:

    i $y = (x + 1)(x + 3)$

    ii $y = (x - 1)(x - 3)$

    iii $y = (x + 1)(x - 3)$

    iv $y = (x - 1)(x + 3)$

  b What do you notice about your graphs?

13 a Give four possible equations for graph A.

  b Give four possible equations for graph B.

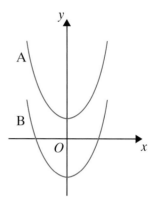

Explore

  ◎ On the same axes draw the graphs of some functions of the form $y = ax^2$ using both positive and negative values of a (for example, $y = x^2$, $y = 3x^2$ and $y = -3x^2$)

    Find out how the graph varies as a varies

  ◎ On the same axes draw the graphs of some functions of the form $y = x^2 + c$ using both positive and negative values of c (for example, $y = x^2 + 1$, $y = x^2 - 2$); find out how the graph varies as c varies

      Investigate further

Learn 2   Graphs of harder quadratic functions

Examples:

  a Draw the graph of $y = 5 + 3x - x^2$ for values of x from -2 to 5.

  b Use the graph to find the solutions of $5 + 3x - x^2 = 0$

  c i Find the x-coordinates of the points where the curve crosses the line $y = 6$

    ii Write a quadratic equation whose solutions are the answers to part **i**.

a The table below gives values for this function.

x	−2	−1	0	1	2	3	4	5
y	−5	1	5	7	7	5	1	−5

The highest value of y in the table is 7, but the curve rises above this between $x = 1$ and 2; to draw the graph accurately it is useful to work out the value of y when $x = 1.5$
When $x = 1.5$, $y = 5 + 3 \times 1.5 - 1.5^2 = 7.25$

b To find the solutions of $5 + 3x - x^2 = 0$ look at the points on $y = 5 + 3x - x^2$ where $y = 0$ (that is, where the curve crosses the x-axis).
The solutions are $x = -1.2$ and $x = 4.2$ (to 1 d.p.)

Graph of $y = 5 + 3x - x^2$

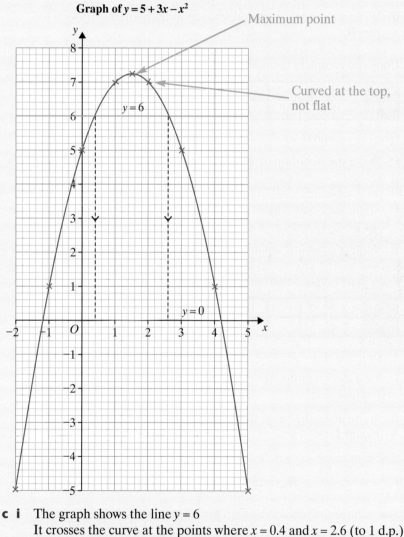

Maximum point

Curved at the top, not flat

$y = 6$

$y = 0$

c i The graph shows the line $y = 6$
It crosses the curve at the points where $x = 0.4$ and $x = 2.6$ (to 1 d.p.)
ii These are the solutions of the equation $5 + 3x - x^2 = 6$
This equation can also be written as:
$3x - x^2 = 1$ or $x^2 - 3x + 1 = 0$

Apply 2

1 a Copy and complete this table for $y = x^2 - 3x$

x	−1	0	1	2	3	4
y	4			−2		4

b Draw the graph of $y = x^2 - 3x$ for values of x from −1 to 4.

c Use your graph to find the solutions of the equation $x^2 - 3x = 0$

d i Draw the line $y = 3$ on your graph.

ii Find the x-coordinates of the points where the line $y = 3$ crosses the curve $y = x^2 - 3x$

iii Write a quadratic equation whose solutions are the answers to part **ii**.

2 a Copy and complete the table for $y = x^2 + 2x + 1$

x	−4	−3	−2	−1	0	1	2
y	9		1	0		4	

b Draw the graph of $y = x^2 + 2x + 1$ for values of x from −4 to 2.

c i Write the x-coordinate of the point where the curve meets the x-axis.

ii Write the quadratic equation whose solution is the answer to part **i**.

d i Write the x-coordinates of the points where the line $y = 8$ crosses the graph of $y = x^2 + 2x + 1$

ii Write a quadratic equation whose solution is the answer to part **i**.

3 a Copy and complete this table for $y = 5 + x - x^2$

x	−3	−2	−1	0	1	2	3	4
y	−7					3		−7

b Draw the graph of $y = 5 + x - x^2$ for values of x from −3 to 4.

c i Find the x-coordinates of the points where the graph crosses the line $y = 0$

ii Write down the quadratic equation whose solution is the answer to part **i**.

d Use your graph to find the solutions of these equations:

i $3 + x - x^2 = 0$ **ii** $x - x^2 = 0$

4 a Draw the graph of $y = 2x^2 - x - 3$ for $-3 \leqslant x \leqslant +4$

b i Write the x-values where the curve crosses the x-axis.

ii Write the quadratic equation whose solutions are the answers to part **i**.

c i Write the x-values where the curve meets the line $y = 10$

ii Write a quadratic equation whose solutions are the answers to part **i**.

5 Get Real!

A ball is thrown vertically upwards into the air.

After t seconds its height above the ground, h metres, is given by the function $h = 1 + 14t - 5t^2$

h metres

a Draw the graph of $h = 1 + 14t - 5t^2$ for $0 \leqslant t \leqslant 3$

b Use your graph to find:

i the height of the ball after 0.25 seconds

ii how long the ball takes to reach a height of 7.2 metres on its way up

iii the maximum height reached by the ball and when this occurs

iv after how long the ball hits the ground.

6 A teacher asks his class to complete a table for $y = 2x^2 - x + 4$

a This is Ella's table but only one of her values for y is correct.

Ella's table

x	−3	−2	−1	0	1	2	3
$2x^2$	36	16	4	0	4	16	36
$-x$	+3	+2	+1	0	−1	−2	−3
$+4$	+4	+4	+4	+4	+4	+4	+4
$y = 2x^2 - x + 4$	43	22	9	4	7	18	37

i Which of Ella's y-values is correct?

ii Explain what Ella has done wrong.

b This is Pete's table.

Pete's table

x	−3	−2	−1	0	1	2	3
$2x^2$	18	8	2	0	2	8	18
$-x$	−3	−2	−1	0	−1	−2	−3
$+4$	+4	+4	+4	+4	+4	+4	+4
$y = 2x^2 - x + 4$	19	10	5	4	5	10	19

i Which of his y-values are correct?

ii Explain any mistakes he has made.

7 a Draw the graph of $y = 5 - 2x - 4x^2$ for $-3 \leqslant x \leqslant +3$

b Use your graph to find the solutions of these equations:

i $5 - 2x - 4x^2 = 0$ **ii** $5 - 2x - 4x^2 = 4$ **iii** $4x^2 + 2x - 7 = 0$

c Luke says that the equation $5 - 2x - 4x^2 = 9$ cannot be solved.

i Is he correct? **ii** Give a reason for your answer.

8 a Draw the graph of $y = 5x^2 + 2x - 4$ for $-3 \leqslant x \leqslant +3$

b i Write the solutions of the equation $5x^2 + 2x = 4$

ii Explain how you found the solution and why your method works.

c i Use your graph to solve the equation $30 - 2x = 5x^2$

ii Explain how you found the solutions and why your method works.

9 a Draw the graph of $y = 4x(x - 2)$ for values of x from −1 to 3.

b For what value of c does the equation $4x^2 - 8x = c$ have just one solution?

c i Describe the values of c for which the equation $4x^2 - 8x = c$ has no solutions.

ii Give a reason for your answer.

191

Explore

◉ Draw the graph of the function $y = ax^2 + bx$ for each set of values of a and b given in this table

Function	a	b	Meets y-axis at	Meets x-axis at	Highest/lowest point
$y = x^2 + 2x$	1	2			
$y = x^2 + 4x$	1	4			
$y = x^2 + 6x$	1	6			
$y = x^2 - 2x$	1	-2			
$y = x^2 - 4x$	1	-4			
$y = x^2 - 6x$	1	-6			
$y = 2x^2 + 2x$	2	2			

◉ Copy and complete the table by entering the coordinates of the points where the curve meets the axes and the coordinates of the highest or lowest point on each curve

Investigate further

Learn 3 Points of intersection of linear and quadratic graphs

Examples:

a i Use a graph to solve the simultaneous equations $y = x^2 + 2x - 4$ and $y = x - 3$

ii How can you tell from the graph that the simultaneous equations $y = x^2 + 2x - 4$ and $y = x - 5$ have no solutions?

i The graph shows the curve $y = x^2 + 2x - 4$ and the lines $y = x - 3$

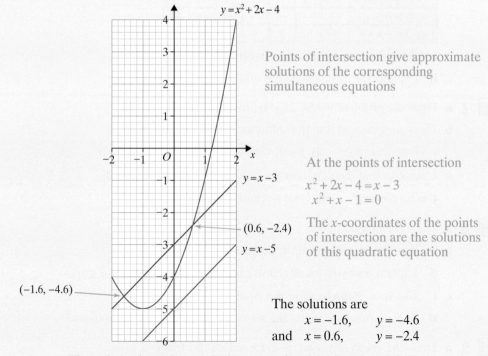

Points of intersection give approximate solutions of the corresponding simultaneous equations

At the points of intersection
$$x^2 + 2x - 4 = x - 3$$
$$x^2 + x - 1 = 0$$

The x-coordinates of the points of intersection are the solutions of this quadratic equation

The solutions are
$$x = -1.6, \quad y = -4.6$$
$$\text{and} \quad x = 0.6, \quad y = -2.4$$

The points where the graphs intersect give the solutions of the **simultaneous equations**.

ii The graph of $y = x - 5$ is parallel to $y = x - 3$ but crosses the y-axis at $(0, -5)$.

It does not meet the curve $y = x^2 + 2x - 4$

So the simultaneous equations $y = x^2 + 2x - 4$ and $y = x - 5$ have no solutions.

The graph of a linear function sometimes meets the graph of a quadratic function at two points, sometimes at only one point and sometimes not at all

b i Complete the table values for $y = x^2 - 2x - 2$

x	−2	−1	0	1	2	3	4
y	6	1			−2	1	

ii Draw the graph of $y = x^2 - 2x - 2$ for values of x between −2 and 4.

iii Write down the solutions of $x^2 - 2x - 2 = 0$

iv By drawing an appropriate linear graph write down the solutions of $x^2 - 3x - 1 = 0$

i Completing the table of values for $y = x^2 - 2x - 2$

x	−2	−1	0	1	2	3	4
y	6	1	−2	−3	−2	1	6

ii

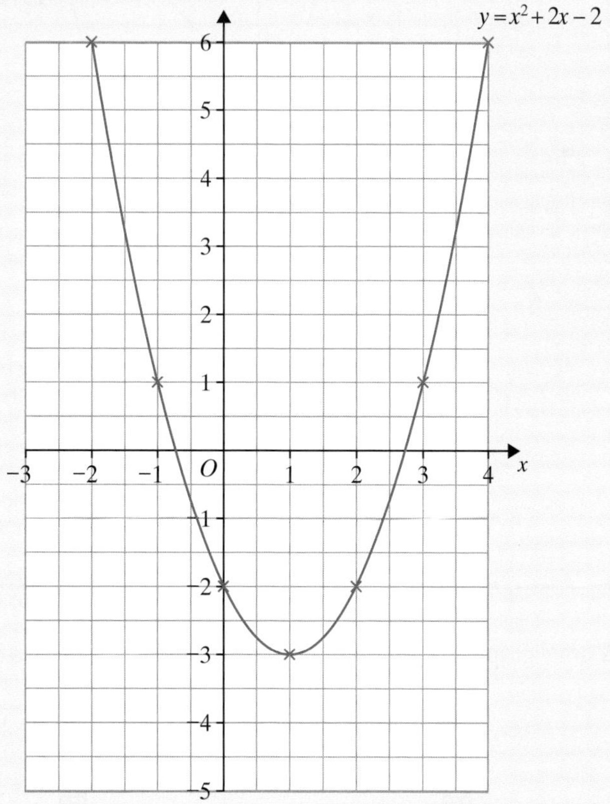

$y = x^2 + 2x - 2$

iii The solutions of $x^2 - 2x - 2 = 0$ are $x = -0.7$ and $x = 2.7$

iv To find the appropriate line:

$$x^2 - 2x - 2 - (x^2 - 3x - 1)$$
$$= x - 1$$

Equation of known line – equation of required line

So we need to draw the line $y = x - 1$

Check:

When $y = x^2 - 2x - 2$ and $y = x - 1$
Then $x^2 - 2x - 2 = x - 1$
$\quad\quad x^2 - 2x - 1 = x$
$\quad\quad x^2 - 3x - 1 = 0$ which is the required equation

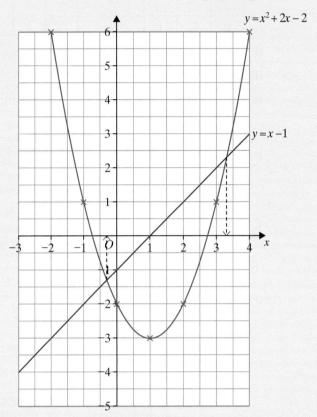

From the graph the solutions of $x^2 - 3x - 1 = 0$ are $x = -0.3$ and $x = 3.3$

Apply 3

In questions **1** to **6** draw graphs for $-4 \leqslant x \leqslant +4$ to solve the simultaneous equations.

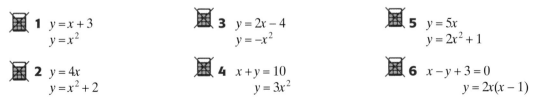

1 $y = x + 3$
$\quad y = x^2$

3 $y = 2x - 4$
$\quad y = -x^2$

5 $y = 5x$
$\quad y = 2x^2 + 1$

2 $y = 4x$
$\quad y = x^2 + 2$

4 $x + y = 10$
$\quad y = 3x^2$

6 $x - y + 3 = 0$
$\quad y = 2x(x - 1)$

7 a On the same axes draw the graphs of $y = 2 + 4x - x^2$ and $y = x - 1$ for $-1 \leqslant x \leqslant +5$

b Use your graph to solve the simultaneous equations $y = 2 + 4x - x^2$ and $y = x - 1$

c Write down and simplify the quadratic equation whose solutions are given by the x-coordinates of the points of intersection of the graphs.

d How can you tell from the graph that the simultaneous equations $y = 2 + 4x - x^2$ and $y = x + 5$ have no solutions?

8 a Draw the graph of $y = 2x^2 - 3x - 2$ for $-1 \leqslant x \leqslant +3$

b By drawing other lines on your graph, use it to find the solutions of these pairs of simultaneous equations.

  i $y = 2x^2 - 3x - 2$ **ii** $y = 2x^2 - 3x - 2$ **iii** $y = 2x^2 - 3x - 2$

       $y = x - 2$ $y = x - 4$ $y = x - 6$

9 a Draw the graph of $y = x^2$ for $-4 \leqslant x \leqslant +4$

b By drawing other lines on your graph, use it to solve these quadratic equations:

  i $x^2 - 3x = 0$ **ii** $x^2 - x - 5 = 0$ **iii** $x^2 + x - 5 = 0$

Remember to write down the equations of your other lines.

10 a Complete the table of values for $y = x^2 - 2x - 1$

x	-2	-1	0	1	2	3	4
y	7		-1		-1		7

b Draw the graph of $y = x^2 - 2x - 1$ for values of x between -2 and 4.

c Write down the solutions of $x^2 - 2x - 1 = 0$

d By drawing an appropriate linear graph write down the solutions of $x^2 - 3x + 1 = 0$

11 Sam says that you can solve the quadratic equation $2x^2 + 3x - 7 = 0$ by finding the points of intersection of the graphs of $y = 2x^2$ and $y = 3x - 7$ Is he correct? Give a reason for your answer.

12 Work out which graph you would draw on the graph of $y = 5x^2$ to solve each of these equations.

a $5x^2 - 4x + 1 = 0$

b $10x^2 - 2x - 3 = 0$

c $6x - 5x^2 = 7$

d $x^2 + 3x = 10$

⊞13 Get Real!

A soft drinks manufacturer finds that when they vary the price of a drink, the amount they sell is related to the price by the function $y = 0.01(x - 100)^2$ where x pence is the price per litre and y is the amount sold in thousands of litres per day.

The amount they are willing to make per day is related to the price by $y = 1.5x - 30$ for $x \geqslant 20$.

a Draw a graph showing $y = 0.01(x - 100)^2$ and $y = 1.5x - 30$ for $0 \leqslant x \leqslant 120$

b Explain why the equation $y = 1.5x - 30$ does not apply in this context when $x < 20$.

c Explain why the equation $y = 0.01(x - 100)^2$ is not likely to apply in this context when $x > 100$.

d Write the coordinates of the point of intersection of the graphs and explain the significance of these values in this context.

Explore

◎ Draw the graph of $y = x^2$ for $-3 \leqslant x \leqslant +3$

◎ On the same axes draw the line $y = 2x + 3$

◎ Write the coordinates of the points of intersection of the curve and line

◎ Write a pair of simultaneous equations whose solutions are given by the x-coordinates of the points of intersection of the line and curve

◎ Write down and simplify the quadratic equation whose solutions are given by the x-coordinates of the points of intersection of the line and curve

◎ Repeat the last three steps using the line $y = 2x + 2$

> **Investigate further**

Quadratic graphs

ASSESS

The following exercise tests your understanding of this chapter, with the questions appearing in order of increasing difficulty.

1 a Taking values of x from -4 to 4, draw tables of values for the functions $y = x^2$, $y = x^2 + 4$ and $y = x^2 - 3$.

b On the same grid, and using the same axes and scales, draw the graphs of these functions.

c Describe the similarities and differences between the graphs.

2 a Taking values of x from -4 to 4, draw tables of values for the functions $y = x^2$, $y = \frac{1}{2}x^2$ and $y = 3x^2$.

b On the same grid, and using the same axes and scales, draw the graphs of these functions.

c Describe the similarities and differences between the graphs.

3 a Copy and complete the table of values for the function $y = 2x^2 - 7$

x	−3	−2	−1	0	1	2	3
y	11		−5	−7		1	

b Draw the graph.

c Use your graph to find:

i the coordinates of the lowest point on the curve

ii the value of y when $x = 1.4$

iii the values of x when $y = 6$

iv the solutions of the equation $2x^2 - 7 = 0$

4 The average safe braking distance for vehicles, d yards, is given by the equation $d = \dfrac{v^2}{50} + \dfrac{v}{3}$, where v is the speed of the vehicle in mph.

a Copy and complete the table of values for the function $d = \dfrac{v^2}{50} + \dfrac{v}{3}$

v (mph)	0	10	20	30	40	50	60	70	80
d (yards)	0	5	15		45	67		121	

b Draw the graph of $d = \dfrac{v^2}{50} + \dfrac{v}{3}$, using v as the horizontal axis.

c Use your graph to find the safe braking distance when the vehicle is travelling at:

i 15 mph   **ii** 45 mph   **iii** 75 mph.

d A driver suddenly sees an obstruction 50 yards ahead. She just stops in time. How fast was she travelling when she first saw it?

5 a Look at the following method to solve the quadratic equation $y^2 - 2y - 8 = 0$

Step 1: Rewrite the equation by adding 8 to both sides $y^2 - 2y = 8$

Step 2: Rewrite the left-hand side as $(y-1)^2 - 1$ $(y-1)^2 - 1 = 8$

Step 3: Tidy up the constants $(y-1)^2 = 9$

Step 4: Take the square root of both sides $y - 1 = \pm 3$

Step 5: Solve the equation $y = 1 \pm 3$ so $y = 4$ or $y = -2$

i In step 2, why is $(y-1)$ squared and not $(y-2)$?

ii In step 2, why is there -1 after the $(y-1)^2$?

b Repeat this method to solve these quadratic equations:

i $x^2 + 2x - 35 = 0$ **iii** $t^2 - 8t + 15 = 0$

ii $d^2 + 14d + 40 = 0$ **iv** $m^2 - m - 72 = 0$

Try some real past exam questions to test your knowledge:

6 a Complete the table of values for $y = 2x^2 - 4x - 1$

x	−2	−1	0	1	2	3
y	15		−1		−1	5

b Draw the graph of $y = 2x^2 - 4x - 1$ for values of x from −2 to 3.

c An approximate solution of the equation $2x^2 - 4x - 1 = 0$ is $x = 2.2$

i Explain how you can find this from the graph.

ii Use your graph to write down another solution of this equation.

Spec A, Higher Paper 1, June 04

7 The grid below shows the graphs of $y = x^2 - x - 6$ and $y = x + 2$

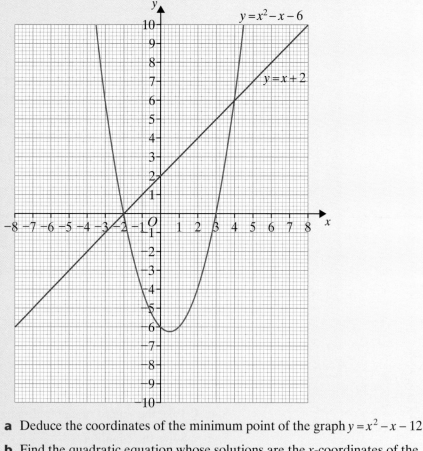

a Deduce the coordinates of the minimum point of the graph $y = x^2 - x - 12$

b Find the quadratic equation whose solutions are the x-coordinates of the points of intersection of $y = x^2 - x - 6$ and $y = x + 2$

Spec A, Higher Paper 2, Nov 03

14 Quadratic functions

OBJECTIVES

B **Examiners would normally expect students who get a B grade to be able to:**

Solve quadratic equations such as $x^2 - 8x + 15 = 0$ by factorisation

A **Examiners would normally expect students who get an A grade also to be able to:**

Solve equations such as $x^2 - 2x - 1 = 0$ by using the quadratic formula

A* **Examiners would normally expect students who get an A* grade also to be able to:**

Solve equations such as $\dfrac{4}{x+2} + \dfrac{3}{2x-1} = 2$

Write quadratic expressions in forms like $(x + a)^2 + b$
(that is, complete the square)

Use completing the square to solve equations and find maximum and minimum values

---

What you should already know ...

- Solve linear equations
- Factorise quadratics
- Simplify surds

VOCABULARY

Quadratic equation – an equation that includes an x^2 term and may also include x terms and constants, for example, $5x^2 + 2x - 4 = 0$; in this example the **coefficient** of x^2 is 5, the coefficient of x is 2 and -4 is the constant term

Solutions of a quadratic equation can be found by graphical methods or (more accurately) by factorising, using the formula or completing the square

Quadratic formula – the solutions (sometimes called the 'roots') of the quadratic equation
$ax^2 + bx + c = 0$ are given by the quadratic formula:
$x = \dfrac{-b \pm \sqrt{b^2 - 4ac}}{2a}$

These solutions are the x-coordinates of the points of intersection of $y = ax^2 + bx + c$ with the x-axis

Discriminant – the expression $b^2 - 4ac$, which is part of the quadratic formula

Completing the square – this refers to writing a quadratic expression in the form $(x + a)^2 + b$ where a and b are positive or negative constants (that is, writing it as a squared term and a constant)

This is useful in solving equations and finding the maximum or minimum value of a quadratic expression

Learn 1   Solving quadratic equations by factorisation

Examples:

a i   Solve $x^2 - x - 6 = 0$

The quadratic equation gives two linear equations

Factorising gives $(x + 2)(x - 3) = 0$
Either $(x + 2) = 0$ or $(x - 3) = 0$ If $a \times b = 0$ then a or $b = 0$
$x = -2$ or $x = 3$

Check each solution in the original equation:
$(-2)^2 - (-2) - 6 = 4 + 2 - 6 = 0$ and $3^2 - 3 - 6 = 9 - 3 - 6 = 0$

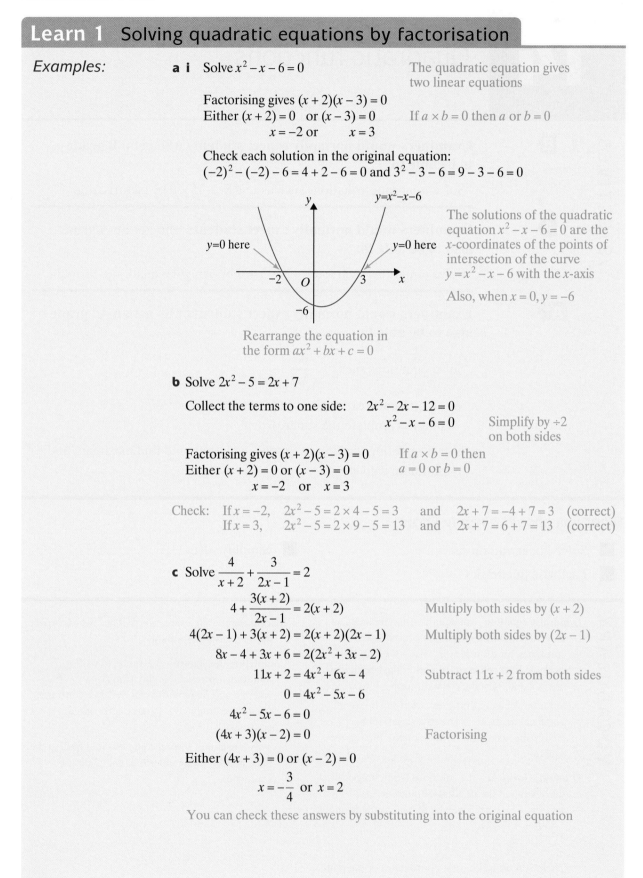

$y = 0$ here $y = 0$ here

The solutions of the quadratic equation $x^2 - x - 6 = 0$ are the x-coordinates of the points of intersection of the curve $y = x^2 - x - 6$ with the x-axis

Also, when $x = 0$, $y = -6$

Rearrange the equation in the form $ax^2 + bx + c = 0$

b Solve $2x^2 - 5 = 2x + 7$

Collect the terms to one side: $2x^2 - 2x - 12 = 0$
$x^2 - x - 6 = 0$ Simplify by $\div 2$ on both sides

Factorising gives $(x + 2)(x - 3) = 0$ If $a \times b = 0$ then
Either $(x + 2) = 0$ or $(x - 3) = 0$ $a = 0$ or $b = 0$
$x = -2$ or $x = 3$

Check: If $x = -2$, $2x^2 - 5 = 2 \times 4 - 5 = 3$ and $2x + 7 = -4 + 7 = 3$ (correct)
         If $x = 3$, $2x^2 - 5 = 2 \times 9 - 5 = 13$ and $2x + 7 = 6 + 7 = 13$ (correct)

c Solve $\dfrac{4}{x + 2} + \dfrac{3}{2x - 1} = 2$

$4 + \dfrac{3(x + 2)}{2x - 1} = 2(x + 2)$ Multiply both sides by $(x + 2)$

$4(2x - 1) + 3(x + 2) = 2(x + 2)(2x - 1)$ Multiply both sides by $(2x - 1)$

$8x - 4 + 3x + 6 = 2(2x^2 + 3x - 2)$

$11x + 2 = 4x^2 + 6x - 4$ Subtract $11x + 2$ from both sides

$0 = 4x^2 - 5x - 6$

$4x^2 - 5x - 6 = 0$

$(4x + 3)(x - 2) = 0$ Factorising

Either $(4x + 3) = 0$ or $(x - 2) = 0$

$x = -\dfrac{3}{4}$ or $x = 2$

You can check these answers by substituting into the original equation

Apply 1

In questions **1** to **12** solve the equations.

1 $x^2 + 4x + 3 = 0$

2 $x^2 + 9x + 14 = 0$

3 $x^2 + 8x + 12 = 0$

4 $x^2 - 7x = 0$

5 $x^2 - 7x + 12 = 0$

6 $x^2 - 16 = 0$

7 $x^2 - 7x - 8 = 0$

8 $x^2 - 8x + 16 = 0$

9 $x^2 + 3x - 18 = 0$

10 $x^2 - 2x = 24$

11 $x^2 - 7 = 6x$

12 $x(x + 2) = 15$

13 Ben is attempting to solve a quadratic equation.
Is Ben correct?
Give a reason for your answer.

$$x^2 - 10x + 24 = 8$$
$$(x - 4)(x - 6) = 8$$
$$x - 4 = 8 \quad \text{or} \quad x - 6 = 8$$
$$x = 12 \quad \text{or} \qquad x = 14$$

14 Get Real!
A rectangular lawn is 4 metres longer than it is wide.
The area of the lawn is 96 m^2.
Using x to represent the width in metres of the lawn:

Area   96 m^2 x

a show that $x^2 + 4x = 96$

b solve this quadratic equation

c find the perimeter of the lawn.

15 Sally thinks of a number, squares it, then adds double the original number.
The result is 80.
Write down a quadratic equation.
Solve the quadratic equation to find
the possible numbers that Sally started with.

16 Adrian says the equation $x^2 + 6x + 9 = 0$ has only one solution.
Is he correct?
Draw a graph and use it to explain your answer.

In questions **17** to **33** solve the equations.

17 $2x^2 + 7x + 3 = 0$

18 $2x^2 - x - 3 = 0$

19 $3x^2 - 14x + 8 = 0$

20 $2x^2 - 18x = 0$

21 $5y^2 + 9y = 2$

22 $4a^2 - 49 = 0$

23 $17p = 7p^2 + 6$

24 $4q^2 + 33q + 35 = 0$

25 $6t^2 = 13t + 5$

26 $(x + 5)(x + 2) = 28$

27 $\dfrac{24}{x + 5} = x$

28 $x = \dfrac{15}{x + 2}$

29 $x = \dfrac{10}{x + 3}$

30 $\dfrac{x + 3}{x} - \dfrac{x + 1}{4} = 1$

31 $\dfrac{3}{x} - \dfrac{2}{x + 2} = 1$

32 $\dfrac{4}{x + 2} + \dfrac{3}{x - 1} = 4$

33 $\dfrac{x}{x + 1} - \dfrac{2}{x - 1} = 2$

201

34 Find the points at which each of these curves meets the *x*-axis.
Illustrate each answer with a sketch of the curve.

   a $y = x^2 - 3x$ **b** $y = x^2 - 7x + 10$ **c** $y = 16 - x^2$ **d** $y = x^2 - 6x + 9$

35 These quadratic functions can all be written as $y = x^2 + bx + c$
In each case find *b* and *c*.

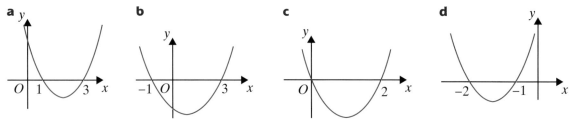

36 Get Real!

An open-topped fish tank is 150 cm long and its width
is 10 cm less than its height.
The capacity of the tank is 45 litres.

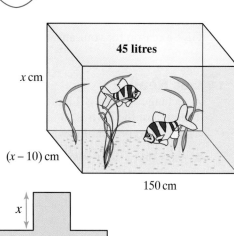

   a Show that $x^2 - 10x - 3000 = 0$

   b Find the width and height of the tank.

   c Calculate the surface area of the outside of the tank,
giving your answer in square metres.

37 This cross is symmetrical.
Its dimensions are given in centimetres.
The area of the cross is 960 cm².

Calculate the value of *x*.

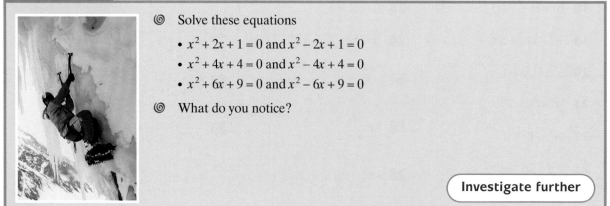

38 The hypotenuse of a right-angled triangle is 15 cm
long. Its other sides differ in length by 3 cm.
Find the perimeter and the area of the triangle.

Explore

◎   Solve these equations

    • $x^2 + 2x + 1 = 0$ and $x^2 - 2x + 1 = 0$

    • $x^2 + 4x + 4 = 0$ and $x^2 - 4x + 4 = 0$

    • $x^2 + 6x + 9 = 0$ and $x^2 - 6x + 9 = 0$

◎   What do you notice?

Investigate further

Learn 2   Solving quadratic equations by using the formula

Example:

Solve $x^2 - 6x + 2 = 0$

$x^2 - 6x + 2 = 0$ cannot be factorised so use the formula

Use the formula $x = \dfrac{-b \pm \sqrt{b^2 - 4ac}}{2a}$ to solve $ax^2 + bx + c = 0$

For this equation $a = 1$, $b = -6$ and $c = 2$

$x = \dfrac{-(-6) \pm \sqrt{(-6)^2 - 4 \times 1 \times 2}}{2 \times 1}$ Substitute the values of a, b and c into the formula

$x = \dfrac{6 \pm \sqrt{(36 - 4 \times 1 \times 2)}}{2 \times 1}$

$x = \dfrac{6 \pm \sqrt{28}}{2}$ Work out the square root and the denominator before splitting into two parts

$x = \dfrac{6 + 5.2915\ldots}{2}$ or $x = \dfrac{6 - 5.2915\ldots}{2}$ Store the value of the square root in the calculator's memory so that you can use it in each part

$x = \dfrac{11.2915\ldots}{2}$ or $x = \dfrac{0.7084\ldots}{2}$

Check:
If $x = 5.65$, $x^2 - 6x + 2 = 5.65^2 - 6 \times 5.65 + 2 = 0.0225$
If $x = 0.35$, $x^2 - 6x + 2 = 0.35^2 - 6 \times 0.35 + 2 = 0.0225$
(not exactly 0 because the solutions were rounded)

$x = 5.65$ (to 2 d.p.) or $x = 0.35$ (to 2 d.p.)

In exam questions you will be asked to give your answers rounded (for example, to 3 s.f. or 2 d.p.) or in surd form (for example, in form $p \pm q\sqrt{r}$)

In this example $x = \dfrac{6 \pm \sqrt{28}}{2} = \dfrac{6 \pm \sqrt{4 \times 7}}{2} = \dfrac{6 \pm 2\sqrt{7}}{2} = 3 \pm \sqrt{7}$

Apply 2

1 Use the formula to find the solutions of these quadratic equations.

   a $x^2 + 4x + 3 = 0$ **b** $x^2 - 7x + 12 = 0$ **c** $2x^2 - x - 3 = 0$

Solve the equations in questions **2** to **16** using the formula.
Give your answers to 2 decimal places.

2 $x^2 + 7x + 5 = 0$

3 $x^2 - 4x + 2 = 0$

4 $x^2 + 5x - 2 = 0$

5 $2x^2 - 7x - 1 = 0$

6 $5x^2 + 2x = 4$

7 $3x^2 = 14x - 5$

8 $7a^2 - 4a = 10$

9 $3x = 20 - x^2$

10 $0.5x^2 - 3.2x + 2 = 0$

11 $4.5y^2 + 24.3y + 3.1 = 0$

12 $35h - 12 = 20h^2$

13 $p(9p - 1) = 4$

14 $r(r + 9) + 5 = 4r^2$

15 $x^2 = (x + 3)(2x + 1)$

16 $\dfrac{3}{2x + 1} = 4 - \dfrac{1}{x - 5}$

17 Find the points at which each of the following curves meet the x-axis.

    a $y = 2x^2 - x - 3$ **b** $y = 5 + x - x^2$ **c** $y = 5x^2 + 2x - 4$

    Interpret your solutions graphically by checking your answers using the graphs you plotted for questions **3**, **4** and **8** in Chapter 13 Apply **2**.

18 Here is Pat's attempt to use the quadratic formula to solve the equation $x^2 - 8 = 3x$
Is Pat correct?
Give a reason for your answer.

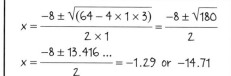

$$x = \frac{-8 \pm \sqrt{(64 - 4 \times 1 \times 3)}}{2 \times 1} = \frac{-8 \pm \sqrt{180}}{2}$$
$$x = \frac{-8 \pm 13.416 \ldots}{2} = -1.29 \text{ or } -14.71$$

19 Get Real!

A rectangular car park is 18 metres longer than it is wide.
The area of the car park is 1200 m².

Using x metres to represent the width of the car park:

    a show that $x^2 + 18x = 1200$

    b find the dimensions of the car park to 1 decimal place.

Area = 1200 m² x

20 Get Real!

The area of a rectangular door is 2.52 m² and its width is 0.9 m less than its height.
Find the dimensions of the door.

21 Get Real!

A picture is 30 cm long and 25 cm wide.
It is mounted in a frame which is x cm wide as shown.
The total area of the frame (including the picture) is 1000 cm².
Find the width of the frame, x cm.

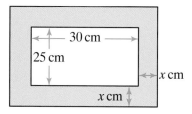

30 cm

25 cm

x cm

x cm

22 A rectangle is 5 cm longer than it is wide. Its diagonals are 20 cm long.
Find the perimeter and area of the rectangle to 3 significant figures.

23 A semicircle has the same area as a square whose sides are 5 cm shorter than the diameter of the semicircle. Find the radius of the semicircle.

24 Solve each of these quadratic equations. Give exact answers using surds.

    a $x^2 + 8x + 5 = 0$ **c** $x^2 - 2x - 9 = 0$ **e** $3x^2 - 8x + 2 = 0$

    b $x^2 - 6x + 4 = 0$ **d** $2x^2 + 4x + 1 = 0$ **f** $2x^2 - 6x - 1 = 0$

25 A girl walks a distance of 20 kilometres and then cycles back.
She takes 1 hour 20 minutes less for the return journey and her average speed is 12 km/h greater than on the first part of her journey.

    a Show that $\dfrac{5}{x} - \dfrac{5}{x + 12} = \dfrac{1}{3}$ where x km/h is her average walking speed.

    b Solve this equation to find her average walking speed.

Explore

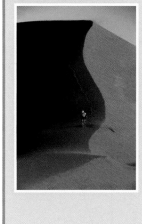

- ◎ In the formula $x = \dfrac{-b \pm \sqrt{b^2 - 4ac}}{2a}$, the expression $b^2 - 4ac$ is called the discriminant

- ◎ Sketch the graph of $y = x^2 + x + 1$
  Find the discriminant of $x^2 + x + 1 = 0$

- ◎ Sketch the graph of $y = x^2 + 2x + 1$
  Find the discriminant of $x^2 + 2x + 1 = 0$

- ◎ Sketch the graph of $y = x^2 + 3x + 1$
  Find the discriminant of $x^2 + 3x + 1 = 0$

- ◎ What do you notice?

Investigate further

Learn 3   Completing the square for a quadratic function

Examples:

a Write $x^2 + 6x + 2$ in the form $(x + a)^2 + b$ where a and b are constants.

> Writing a quadratic in this form is called 'completing the square'

b Write $x^2 + 4x - 7$ in the form $(x + a)^2 + b$ where a and b are constants.

c Use your answer to part **b** to:
  i solve the equation $x^2 + 4x - 7 = 0$
  ii find the lowest point on the graph of $y = x^2 + 4x - 7$ and sketch it.

> This form is useful for solving equations and finding maximum and minimum values

a For $x^2 + 6x + 2$ use the fact that:
$$(x + 3)^2 = x^2 + 6x + 9$$
so $(x + 3)^2 - 9 = x^2 + 6x$
or $x^2 + 6x = (x + 3)^2 - 9$

> The coefficients of x^2 and x are correct so we need to deal with the constants

Substituting
$$x^2 + 6x + 2$$
$$= (x + 3)^2 - 9 + 2$$
$$= (x + 3)^2 - 7$$

> Compare this with $(x + a)^2 b$ to find a and b

so $a = 3$ and $b = -7$

In general $x^2 + bx + c$ can be written $\left(x + \dfrac{b}{2}\right)^2 - \left(\dfrac{b}{2}\right)^2 + c$

> We use the fact that $\left(x + \dfrac{b}{2}\right)^2 = x^2 + bx + \left(\dfrac{b}{2}\right)^2$

> So $x^2 + bx = \left(x + \dfrac{b}{2}\right)^2 - \left(\dfrac{b}{2}\right)^2$

b $x^2 + 4x - 7$
$= (x + 2)^2 - 2^2 - 7$
$= (x + 2)^2 - 4 - 7$
$= (x + 2)^2 - 11$

> Use the fact that $x^2 + bx + c$ can be written as $\left(x + \dfrac{b}{2}\right)^2 - \left(\dfrac{b}{2}\right)^2 + c$

c i Rewriting $\qquad x^2 + 4x - 7 = 0$
gives $\qquad (x+2)^2 - 11 = 0$

so $\quad (x+2)^2 = 11$ — Square rooting both sides
$\quad\quad x + 2 = \pm\sqrt{11}$ — Subtracting 2 from both sides
$\quad\quad\quad x = -2 \pm \sqrt{11}$
$\quad\quad\quad x = -2 \pm 3.3166\ldots$
$\quad\quad\quad x = -2 + 3.3166\ldots$ gives $x = 1.32$ (to 2 d.p.)
or $x = -2 - 3.3166\ldots$ gives $x = -5.32$ (to 2 d.p.)

Remember that you should check the solutions in the equation

ii $y = x^2 + 4x - 7 = (x+2)^2 - 11$
Squared terms are never negative, so the lowest value of $(x+2)^2$ is 0 when $x = -2$

So $y = x^2 + 4x - 7$ has a minimum value of -11 when $x = -2$

$(-2, -11)$ is the lowest point on the graph of $y = x^2 + 4x - 7$

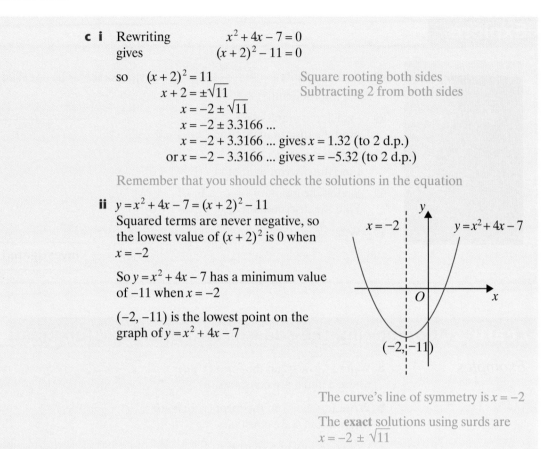

The curve's line of symmetry is $x = -2$

The **exact** solutions using surds are $x = -2 \pm \sqrt{11}$

Apply 3

In questions **1** to **3**, write the equation in the form $(x+a)^2 + b = 0$ and hence find the solutions of each equation to 2 decimal places.

1 $x^2 + 2x - 1 = 0$ $\qquad$ **2** $x^2 + 6x - 5 = 0$ $\qquad$ **3** $x^2 + 8x + 1 = 0$

In questions **4** to **6**, write the equation in the form $(x-a)^2 + b = 0$ and hence find the solutions of each equation to 2 decimal places.

4 $x^2 - 10x - 5 = 0$ $\qquad$ **5** $x^2 - 6x + 1 = 0$ $\qquad$ **6** $x^2 - 5x - 9 = 0$

In questions **7** to **9**, write each equation in the form $(x+a)^2 = b$ and hence find the solutions, giving your answers in surd form.

7 $x^2 + 8x = 1$ $\qquad$ **8** $x^2 + 6x = 9$ $\qquad$ **9** $x^2 + 3x = 1$

In questions **10** to **12**, write each equation in the form $(x-a)^2 = b$ and hence find the solutions, giving your answers in surd form.

10 $x^2 - 4x = 8$ $\qquad$ **11** $x^2 - 2x = 2$ $\qquad$ **12** $x^2 - x = 7$

In questions **13** to **18**:

i write the function in the form $y = (x + a)^2 + b$ where a and b are positive or negative constants

ii find the coordinates of the lowest point on the curve

iii sketch the curve and show its line of symmetry.

13 $y = x^2 + 2x + 7$ **16** $y = x^2 + 8x - 1$

14 $y = x^2 - 4x - 6$ **17** $y = x^2 - 5x$

15 $y = x^2 - 6x + 3$ **18** $y = x^2 - 3x + 5$

19 Greg has been asked to write $x^2 + 4x + 1$ in the form $(x + a)^2 + b$
Here is his working. Is Greg correct?
Give reasons for your answer.

> $(x + a)^2 + b = x^2 + a^2 + b$
> If this $= x^2 + 4x + 1$
> then $b = 1$
> and $a^2 = 4$ so $a = 2$

20 a Find a and b such that $2x^2 - 12x + 11 = 2(x - a)^2 + b$ where a and b are positive or negative.

b Use your answer to part **a** to:

i solve the equation $2x^2 - 12x + 11 = 0$

ii find the lowest point on the graph of $y = 2x^2 - 12x + 11$ and sketch it.

21 a Find a and b such that $2 - 8x - x^2 = b - (x + a)^2$

b Use your answer to part **a** to:

i solve the equation $2 - 8x - x^2 = 0$

ii find the highest point on the graph of $y = 2 - 8x - x^2$ and sketch it.

22 A right-angled triangle has sides of length x cm, $(x - 3)$ cm and $(x - 5)$ cm as shown.

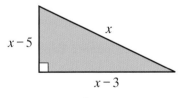

a Show that $x^2 - 16x + 34 = 0$

b Find the solutions of this equation by completing the square.

c Write the lengths of the sides of the triangle to the nearest millimetre.

23 Get Real!
The sketch shows the cross-section of a storage tank with its dimensions given in metres.

a If the area of the cross-section is 4.39 m^2, show that $(x + 0.1)^2 = 4.4$

b Hence find the value of x to 1 decimal place.

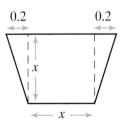

Quadratic functions

The following exercise tests your understanding of this chapter, with the questions appearing in order of increasing difficulty.

1 Solve these quadratic equations by factorisation.

  a $a^2 - 7a + 12 = 0$

  b $b^2 + 2b - 15 = 0$

  c $c^2 - 4c - 12 = 0$

  d $d^2 + 11d + 28 = 0$

2 Rearrange the equation $2x - \dfrac{15}{x} + 7 = 0$ into a quadratic form and solve it by factorisation.

3 Use the formula to solve these quadratic equations. Give your answers to 2 decimal places.

  a $x^2 + 3x - 7 = 0$

  b $3x^2 - 12x + 10 = 0$

Try some real past exam questions to test your knowledge:

4 $x^2 - 6x + 13 = (x - a)^2 + b$

  a Find the values of a and b.

  b Hence find the minimum value of $x^2 - 6x + 13$.

Spec B, Module 5 Paper 2, June 03

5 Solve the equation $\dfrac{x}{x + 1} - \dfrac{2}{x - 1} = 1$

Spec B, Module 5 Paper 2, June 03

15 Inequalities and simultaneous equations

OBJECTIVES

C   **Examiners would normally expect students who get a C grade to be able to:**

Solve inequalities such as $3x < 9$ and $12 \leqslant 3n < 20$

Solve linear inequalities such as $4x - 3 < 10$ and $4x < 2x + 7$

Represent sets of solutions on the number line

B   **Examiners would normally expect students who get a B grade also to be able to:**

Solve linear inequalities such as $x + 13 > 5x - 3$

Solve a set of linear inequalities in two variables and represent the solution as a region of a graph

Solve a pair of simultaneous equations in two unknowns, such as $2x + y = 5$ and $3x - 2y = 4$

Know that each equation can be represented by a line on a graph and that the point of intersection of the lines is the solution

A   **Examiners would normally expect students who get an A grade also to be able to:**

Solve a pair of simultaneous equations where one is linear and one is non-linear, such as $y = 3x - 2$ and $y = x^2$

A*   **Examiners would normally expect students who get an A* grade also to be able to:**

Solve a pair of simultaneous equations where one is linear and one is non-linear, such as $x + 5y = 13$ and $x^2 + y^2 = 13$

What you should already know ...

- Inequality signs
- Draw a graph from an equation
- Find the equation of a straight line
- Collect like terms
- Multiply out brackets

- Cancel fractions
- Add and subtract fractions
- Solve equations
- Solve quadratic equations by factorising and by using the quadratic formula

VOCABULARY

Integer – any positive or negative whole number or zero, for example, $-2, -1, 0, 1, 2 \ldots$

Inequality – statements such as $a \neq b$, $a \leqslant b$ or $a > b$ are inequalities

Inequality signs – $<$ means less than, $\leqslant$ means less than or equal to, $>$ means greater than, $\geqslant$ means greater than or equal to

Number line – a line where numbers are represented by points upon it; simple inequalities can be shown on a number line

Real number – any rational or irrational number; all real numbers can be represented on the number line

$-5 \quad 42 \quad 0 \quad \pi \quad \sqrt{2} \quad \frac{4}{9}$

Simultaneous equations – two equations that apply simultaneously to given variables; the solution to the simultaneous equations is the pair of values for the variables that satisfies both equations; the *graphical* solution to simultaneous equations is a point where the lines representing the equations *intersect*

Unknown – the letter in an equation such as x or y

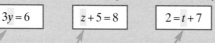

$\boxed{3y = 6}$ $\boxed{z + 5 = 8}$ $\boxed{2 = t + 7}$

y is the unknown z is the unknown t is the unknown

Linear equation – an equation where the highest power of the variable is 1; for example, $3x + 2 = 7$ is a linear equation but $3x^2 + 2 = 7$ is not

Non-linear equation – an equation that cannot be represented by a straight line graph, for example, $3x^2 + 2 = 7$, $x^2 + y^2 = 25$ and $xy = 10$ are non-linear equations

Solution – the value of the unknown in an equation, for example, the solution of the equation $3y = 6$ is $y = 2$

Solve – when you solve an equation you find the solution, in an equation or expression

Coefficient – the number (with its sign) in front of the letter representing the unknown, for example:

$\boxed{4p - 5}$ $\boxed{2 - 3p^2}$

4 is the coefficient of p -3 is the coefficient of p^2

Brackets – these show that the terms inside should be treated alike, for example,

$2(3x + 5) = 2 \times 3x + 2 \times 5 = 6x + 10$

Simplify – to make simpler by collecting like terms

Substitute – find the value of an expression when the variable is given a value, for example, when $x = 4$, the expression $3x + 2 = 3 \times 4 + 2 = 14$

Quadratic equation – an equation that includes an x^2 term and may also include x terms and constants, for example, $5x^2 + 2x - 4 = 0$; in this example the **coefficient** of x^2 is 5, the coefficient of x is 2 and -4 is the constant term

Solutions of a quadratic equation can be found by graphical methods or (more accurately) by factorising, using the formula or completing the square

Learn 1   Inequalities and the number line

Examples:

a Draw number lines to show these inequalities.

i $x < 2$ **ii** $x \geqslant -3$ **iii** $-6 < x \leqslant -3$ **iv** $x \leqslant -2$ or $x > 1$

i

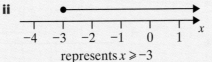

represents $x < 2$

An **open** circle shows that x can be very close to 2 but not equal to 2

If x is an integer, it could be 1, 0, –1, –2, –3, ... (but not 2).
If x is any real number, it could be any number less than 2.

ii

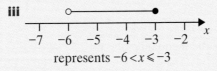

represents $x \geqslant -3$

A **closed** circle shows that x can equal –3

If x is an integer, it could be –3, –2, –1, 0, 1, 2, ...
If x is any real number, it could be any number greater than or equal to –3.

iii

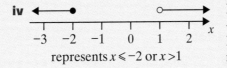

represents $-6 < x \leqslant -3$

If x is an integer, it could only be –5, –4 or –3 (but not –6).
If x is any real number, it could be any number greater than –6 **and** less than or equal to –3.

iv

represents $x \leqslant -2$ or $x > 1$

If x is an integer, it could be –2, –3, –4 ... or 2, 3, 4, ... (but not –1, 0 or 1).
If x is any real number, it could be any number less than or equal to –2 **or** any number greater than 1.

b Solve $-5 < 3x + 4 \leqslant 10$

You can solve inequalities using the same methods as for solving equations.

First, split the inequality into two parts.

$-5 < 3x + 4$ and $3x + 4 \leqslant 10$
  $-9 < 3x$ and $3x \leqslant 6$ Subtract 4 from both sides
  $-3 < x$ and $x \leqslant 2$ Divide both sides by 3

Add or subtract the same number from both sides and the inequality is still valid

Finally, put the inequalities back together: so $-3 < x \leqslant 2$

If the question had said that x is an integer, the possible values are –2, –1, 0, 1, 2

Multiply or divide both sides by the same **positive** number and the inequality stays valid

c Find the largest integer that satisfies $x + 13 > 5x - 3$

$x + 13 > 5x - 3$
  $13 > 4x - 3$ Subtract x from both sides
  $16 > 4x$ Add 3 to both sides
  $4 > x$ Divide both sides by 4
  $x < 4$

However, if you multiply or divide both sides by a **negative** number, the inequality is reversed

x is less than 4, so x can be any integer less than 4, but not 4

The largest integer that satisfies the inequality is 3.

Apply 1

1 Show each of the following inequalities on a number line:

   a $x < -2$ **d** $x \leqslant 4$ **g** $x < -2$ or $x > 1$

   b $x < -3$ **e** $-2 \leqslant x \leqslant 3$ **h** $x \leqslant 2$ or $x > 5$

   c $x \geqslant -1$ **f** $-4 < x < 0$ **i** $-2 < x < 2$

2 Write the inequalities shown on these diagrams.

   a

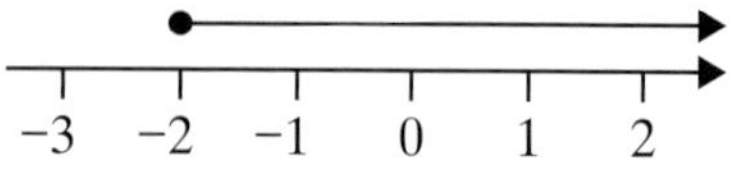

   b

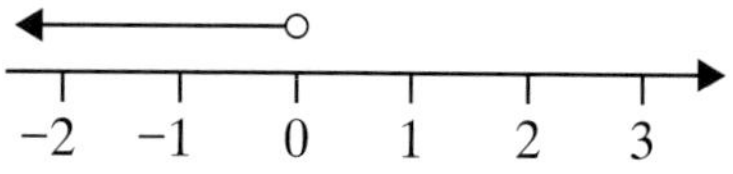

   c

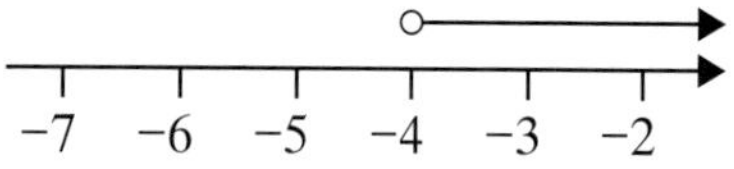

   d

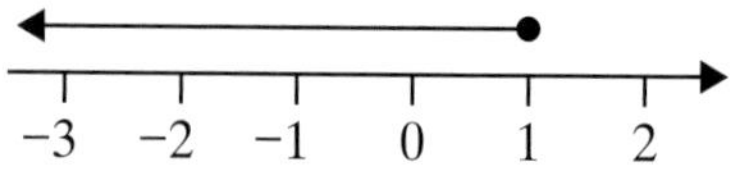

   e

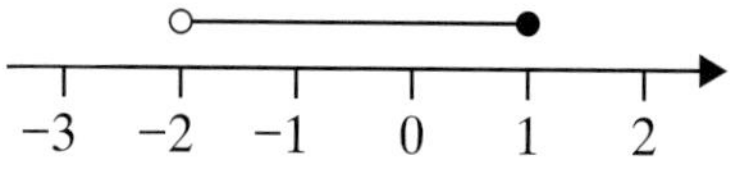

   f

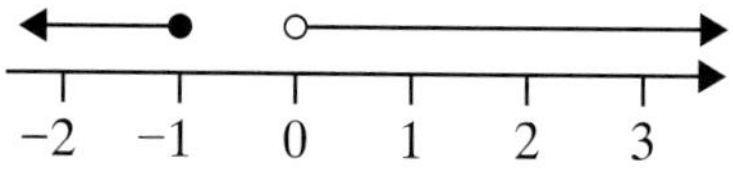

3 Oliver says that this diagram shows the inequality $-4 < x < 2$.
Julia says that he is wrong, it shows $-4 \leqslant x \leqslant 2$.
Do you agree with Oliver or Julia?
Give a reason for your answer.

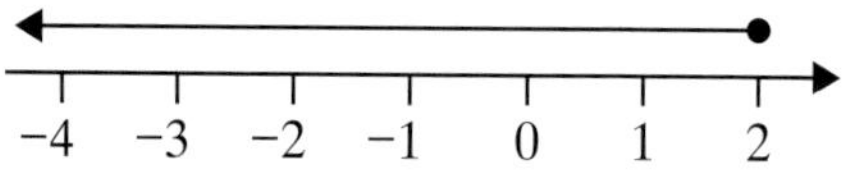

4 List all the integer values of n such that:

   a $-8 \leqslant 4n < 15$ **b** $-3 < 2n \leqslant 12$ **c** $-5 \leqslant 2n - 1 < 6$

5 Solve these linear inequalities.

   a $4x - 1 \geqslant 3$ **d** $x < 4x - 9$ **g** $4 > 7 - x$

   b $2x + 16 \leqslant 29$ **e** $3x - 7 > 8x + 8$ **h** $12 + 2x < 6 - x$

   c $5x + 10 < 0$ **f** $4(x + 3) \leqslant 3(x - 2)$ **i** $6 - 4x \geqslant 3 - x$

6 Given that $5y > 2$ and $\frac{1}{2}y \leqslant 3\frac{1}{2}$ and that y is an integer,
find the possible values of y.

7 Find the smallest integer that satisfies:

   a $0 > 4 - x$ **c** $2x + 8 \geqslant 0$

   b $3(2x - 10) \geqslant 2$ **d** $6 \leqslant 5(2x + 7)$

8 Find the largest integer that satisfies:

   a $9 - 2x \geqslant 5$ **c** $6 - 3x \geqslant 10$

   b $4(3x + 9) < 50$ **d** $7(2x - 5) < 4x + 5$

9 Find all the possible pairs of positive integers, x and y,
such that $2x + 3y \leqslant 9$.

10 Solve these linear inequalities.

a $8 \geqslant 12 - \dfrac{x}{4}$

c $\dfrac{3x + 8}{4} > 1$

e $\dfrac{x}{8} + 5 \geqslant 4 - \dfrac{x}{4}$

b $\dfrac{2x - 3}{5} < 6$

d $\dfrac{x}{3} - \dfrac{x}{4} \leqslant -2$

f $\dfrac{3 - 2x}{4} < x + 3$

11 Get Real!

Tolu is saving up for a new bike that costs £99.
She has £20 in her account at the moment.
Write an inequality and solve it to find the least number of pounds that
Tolu must save every month for the next five months if she is to have
enough money for the bike.

12 Two integers x and y are such that $2 \leqslant x < 7$ and $-4 \leqslant y \leqslant 1$.

a What is the largest value of y^2?

b What is the smallest value of xy?

c If $y^2 = 4$, what is the value of y?

13 Get Real!

Class 11Y want to know how old their teacher is.
Mrs Hirst gives them a clue.
She says: 'In 10 years time I will be **less than** double
the age I was 10 years ago'.
Can you write an inequality and solve it to find the
youngest age that Mrs Hirst could be?

Explore

- ◎ Think of two numbers that satisfy the inequality $x^2 < 9$
- ◎ Find the two solutions of the equation $x^2 = 9$
- ◎ Show the set of solutions of $x^2 < 9$ on a number line
- ◎ Investigate which values of x satisfy the inequality $x^2 > 9$
- ◎ Where does this set of solutions fit on the number line?
- ◎ Now look at the solutions of $x^2 = 4$

Investigate further

Learn 2　Inequalities and regions

Examples:

a Show the regions defined by these inequalities.

You can use a graph to show an inequality

i $x \geqslant 2$　　**ii** $y < 1$　　**iii** $x + y > 2$

The inequality is $\geqslant$ so use a solid line

i　$x \geqslant 2$

Plot the **boundary line** of the inequality: change the inequality sign to an equals sign and plot the line.
The solid line is $x = 2$
The shaded region is $x \geqslant 2$

Test a point to see which side of the line you want, for example, $(0, 0)$: when $x = 0$, $0 \leqslant 2$, so the inequality is not satisfied and $(0, 0)$ is not in the region

The inequality is $<$ so use a broken line

ii　$y < 1$

The dotted line is $y = 1$
The shaded region is $y < 1$

Test $(0, 0)$ in the inequality: when $y = 0$, $0 < 1$, so the inequality is satisfied and $(0, 0)$ is in the region

The inequality is $>$ so use a broken line

iii $x + y > 2$

The dotted line is $x + y = 2$
The shaded region is $x + y > 2$

Test $(0, 0)$ in the inequality: $0 + 0 < 2$, so the inequality is not satisfied and $(0, 0)$ is not in the region

In the exam it is better if you shade using hatching because colour does not always show clearly, for example, $x \leqslant 1$

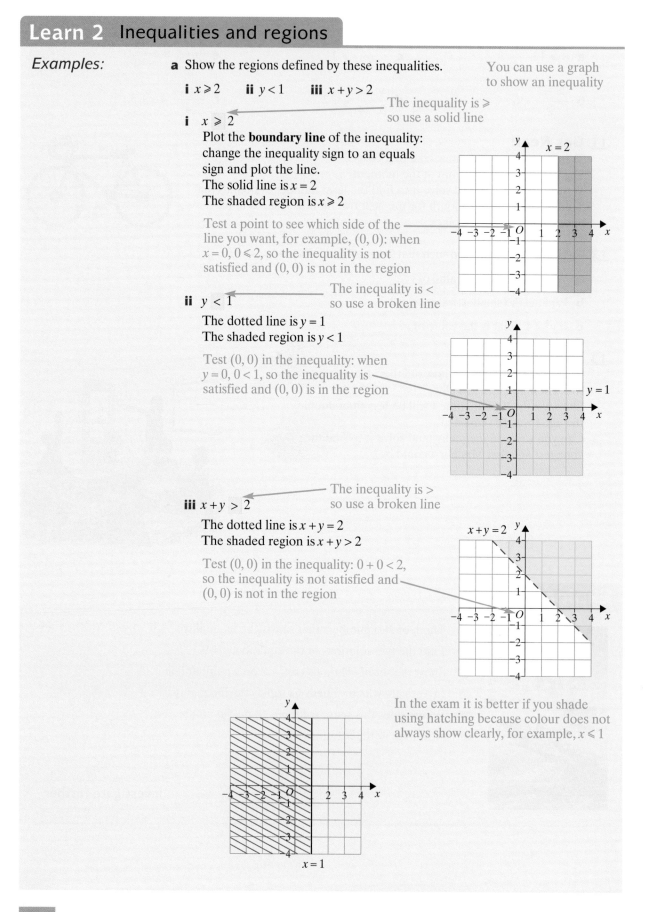

b Show the region defined by the inequalities $x > 1$, $y \geqslant 0$ and $x + y \leqslant 3$.
If x and y are integers, which points satisfy all three inequalities?

Draw the line $x = 1$ (broken because the inequality is >) and check to see which side is $x > 1$
(Test the point $(0, 0)$: when $x = 0$, $0 < 1$ so $(0, 0)$ isn't in the region)
Shade the side that **is not** in the region

The line $y = 0$ is already drawn
It is the x-axis
Leave it solid because the inequality is $\geqslant$

The region above the x-axis is where $y > 0$, so shade the side that **is not** in the region, that is the region below the x-axis

Draw the line $x + y = 3$ (solid because the inequality is $\leqslant$) and check to see which side is $x + y \leqslant 3$

Required region

It is a good idea to clearly identify the required region

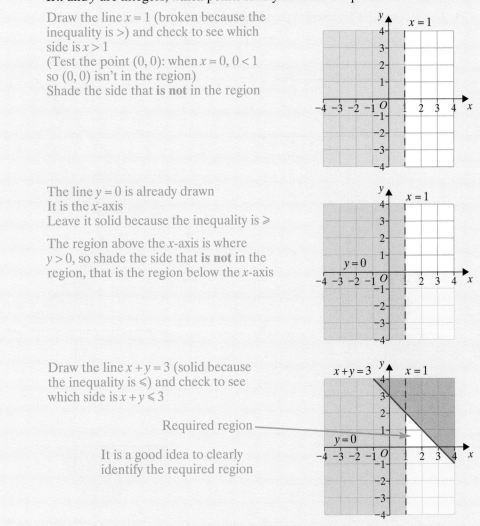

The required region is the **unshaded** triangle.

If x and y are integers, then the points $(2, 0)$, $(3, 0)$ and $(2, 1)$ are in the region.

$(1, 0)$, $(1, 1)$ and $(1, 2)$ are on a dotted line so they are not in the region

Apply 2

1 Use inequalities to describe the **shaded** regions.

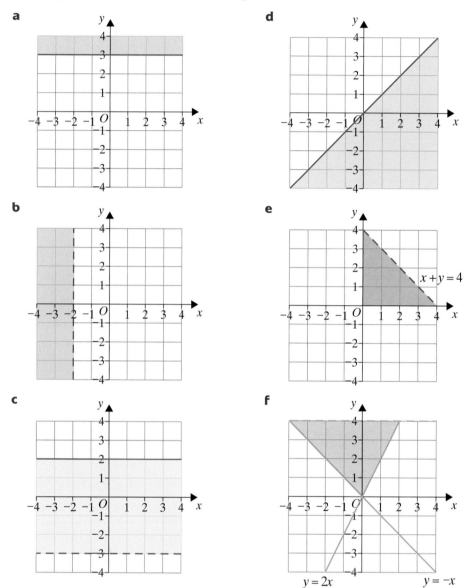

2 Draw, on separate diagrams, the regions defined by these inequalities.
Use hatched shading to show your regions clearly.

a $x \geqslant 1$

b $y \leqslant -1$

c $x > -3$

d $-2 < x \leqslant 1$

e $y \geqslant 2x$

f $-1 \leqslant y < 3$ and $-2 < x \leqslant 0$

3 a Draw x- and y-axes from -2 to 5. Draw the lines $y = 1$, $y = 2x$ and $x + y = 5$.

b Identify the region where $y > 1$, $y < 2x$ and $x + y < 5$.

c Should the boundary lines of your region be solid or dotted?
Give a reason for your answer.

4 Ryan says that the graph shows the inequality $y < 2$.
Charlie says it shows $y \leq 2$.
Who is correct?
Give a reason for your answer.

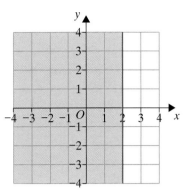

5 Match these inequalities with the **unshaded** regions of the graphs.

a $y < 3 - x, y \geq 0, x \geq 0$ **b** $y > x - 3, y \leq 0, x \geq 0$

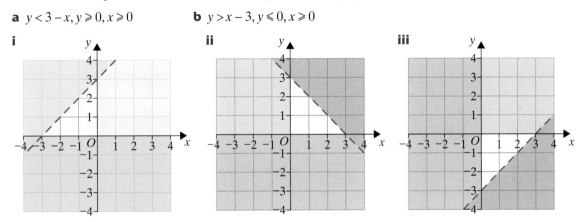

c Describe the unshaded region of the remaining graph.

6 Find the equations of the straight lines on these graphs.
Use inequalities to describe the **shaded** regions.

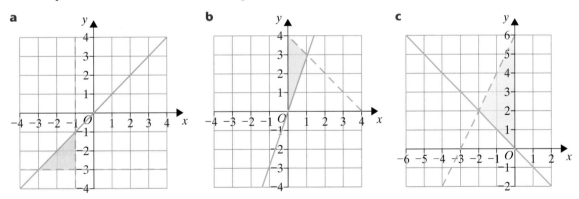

7 a Draw the graph of $y = 5 - 2x$.
Shade with hatching the region where $2x + y \leq 5$.

  b List the coordinates of all the points where $2x + y \leq 5$ and x and y are positive integers.

8 a Draw and clearly label the region defined by the three inequalities $x \geq 1$, $y < 4 - x$ and $x + 2y \geq 2$.

  b List all the points in the region whose coordinates are integers.

9 **a** Draw the region defined by the three inequalities $y \geqslant 0$, $y \geqslant 2x - 3$ and $3y \leqslant 12 - 4x$.

  b If x and y are integers in the region and $y = 3x - 1$, find x and y.

10

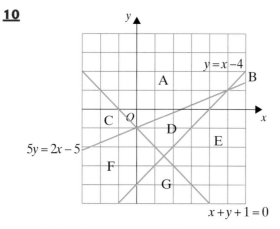

a In the diagram state which region bounded by blue lines is described by:

  i $y \geqslant x - 4$, $5y \leqslant 2x - 5$, $x + y + 1 \leqslant 0$

  ii $y \leqslant x - 4$, $5y \geqslant 2x - 5$

  iii $y \geqslant x - 4$, $5y \leqslant 2x - 5$, $x + y + 1 \geqslant 0$

b Describe with inequalities these regions bounded by blue lines.

  i E

  ii G

  iii A + C

  iv D + E

11 Get Real!

At the start of the week Jonathan had £7 in his pocket.
He remembers that he only had £1 and £2 coins.
He has lost some money and now has x £1 coins and y £2 coins.

a Write three inequalities.

b Draw a graph of your inequalities and use the coordinates of the region to write down all the possible combinations of coins in Jonathan's pocket.

Explore

- ◎ Graphs of inequalities are used in **linear programming** to find the best solution to a problem where several conditions need to be satisfied at the same time

- ◎ Here is an example of a real-life problem that you can solve:
  - Sue makes home-made cards to sell; she makes birthday cards for boys and girls
  - Her 'birthday girl' cards are more popular so she always makes at least twice as many of these
  - The most she can make in a week is 14 cards altogether
  - She always makes at least three 'birthday boy' cards

- ◎ If Sue makes x 'birthday boy' cards and y 'birthday girl' cards in a week, write down three inequalities for x and y

- ◎ Draw a graph of your three inequalities and identify the **feasible region** – the region where all the inequalities are satisfied

- ◎ List the possible combinations of cards she can make that satisfy all the conditions

- ◎ Sue makes £1.80 profit on each of the 'birthday boy' cards and £1.50 profit on each of the 'birthday girl' cards; what combination should she make to get the most profit?

- ◎ Make up your own problem with conditions to be satisfied

Investigate further

Learn 3 Solving simultaneous equations

Examples:

a Solve this pair of simultaneous equations.

$$4x + 3y = 13$$
$$2x - 3y = 11$$

The coefficients of y are matching numbers with opposite signs, so add the equations to eliminate y

$$4x + 3y = 13 \quad ①$$
$$2x - 3y = 11 \quad ②$$

Adding ① + ② gives:

$$4x + 3y = 13$$
$$+2x - 3y = 11$$
$$\overline{\quad 6x \qquad = 24}$$

Divide by 6 on both sides

$$+3y + -3y = 0$$

Solving to find x: $x = 4$

Now substituting $x = 4$ into ①:

$$4 \times 4 + 3y = 13$$
$$16 + 3y = 13$$
$$3y = -3 \qquad \text{Subtract 16 on both sides}$$
$$y = -1 \qquad \text{Divide by 3 on both sides}$$

Checking the solution in ② gives:

$$2 \times 4 - 3 \times -1 = 8 + 3 = 11 \quad ✓$$

The solution of two simultaneous equations is the point of intersection of the two lines; so the lines $4x + 3y = 13$ and $2x - 3y = 11$ intersect at the point $(4, -1)$

So the solution is $x = 4$ and $y = -1$

219

b Solve the equations:

$6x + 2y = -10$
$2x + 4y = -10$

$6x + 2y = -10$ ①   Make the coefficients of y match by multiplying
$2x + 4y = -10$ ②   the first equation by 2 (or you could multiply
the second equation by 3 to make the
coefficients of x match)

① × 2 $12x + 4y = -20$ ③   The coefficients of y are now matching
       $2x + 4y = -10$ ②   numbers with the same sign, so subtract
the equations to eliminate y

Subtracting ③ – ② gives: $12x + 4y = -20$ Be careful!
           $- \quad (2x + 4y = -10)$ $-20--10 = -20+10 = -10$
             $\overline{10x \qquad = -10}$ Dividing by 10 on both sides

Solving to find x: $x = -1$

Now substituting $x = -1$ ino ①:

$6 \times -1 + 2y = -10$
$-6 + 2y = -10$ Adding 6 to both sides
$2y = -4$ Dividing both sides by 2
$y = -2$

Checking the solution in ② gives:

$2 \times -1 + 4 \times -2 = -2 + -8 = -10$ ✓

So the solution is $x = -1$ and $y = -2$

c Solve the equations:

$4x - 6y = 5$
$10x + 4y = 3$

Make the coefficients of y match (with opposite
signs) by finding a common multiple of 6 and 4,
for example, 12

$4x - 6y = 5$ ①   Multiply the first equation by 2 and the second
$10x + 4y = 3$ ②   equation by 3

① × 2 $8x - 12y = 10$ ③   The coefficients of y are now matching
② × 3 $30x + 12y = 9$ ④   numbers with opposite signs, so add
the equations to eliminate y

Adding ③ + ④ gives: $8x - 12y = 10$
          $+ \quad 30x + 12y = 9$
            $\overline{38x \qquad = 19}$

Solving to find x: $x = \frac{1}{2}$

Now substituting $x = \frac{1}{2}$ into ①:

$4 \times \frac{1}{2} - 6y = 5$
$2 - 6y = 5$ Subtract 2 from both sides
$-6y = 3$ Dividing by -6 on both sides
$y = -\frac{1}{2}$

Checking the solution in ② gives:

$10 \times \frac{1}{2} + 4 \times -\frac{1}{2} = 5 + -2 = 3$ ✓

So the solution is $x = \frac{1}{2}$ and $y = -\frac{1}{2}$

Apply 3

1 Solve the following pairs of simultaneous equations.

a $3x - y = 10$
$x + y = -2$

d $3s + 4t = 25$
$5s - 4t = -33$

g $5c + 4d = 6$
$-5c + 3d = 22$

b $2a + 3b = 13$
$a + 3b = 11$

e $4x - 3y = 0$
$2x + 3y = 9$

h $2x + 3y = 0$
$2x - 5y = -4$

c $5x + y = 14$
$3x + y = 10$

f $m + 3n = -4$
$m - n = 0$

2 Get Real!

The sum of the ages of Jack's aunt and uncle is 88 years.
The difference in their ages is 16 years.
If Jack's aunt is older than his uncle, what are their ages?

3 Get Real!

Carla is making a new rectangular fishpond for her garden.
She has enough stones to go around a pond of perimeter 10.5 metres.
She would like the longer sides to be twice the length of the shorter sides.
By calling the length of the longer sides y and the shorter sides x, form
two equations in x and y and solve them simultaneously to find the
measurements of Carla's fishpond.

4 Solve the following pairs of simultaneous equations.

a $5x + 3y = 27$
$2x + y = 10$

d $2e + 3f = 2$
$e + 6f = 4$

g $3x - 4y = 5$
$2x + y = -4$

b $a + 3b = 21$
$2a - b = 14$

e $4x + 3y = 1$
$12x = 14 + 2y$

h $3j + 2k = 5$
$5j - 6k = 27$

c $9x + 7y = 10$
$3x + y = 2$

f $2p + 3q = 5$
$6q = 12 - 5p$

5 Get Real!

In a kindergarten there are x bicycles and y tricycles.
There are enough for each of the 120 children to have a ride
at the same time.
One of the children counts all the wheels and finds that there
are 315 in total.
How many bicycles and how many tricycles are there?

6 Get Real!

Geoff sells two kinds of Christmas tree decorations
on his market stall.
Angels cost £3 and Santas cost £2.
On one day he sells 75 decorations for £183.
How many of each does Geoff sell?

7 Write three pairs of simultaneous equations with the solution
$x = 3, y = 2$.

8 Find a and b if $a + 3b = 4a - 2b = 7$

9 Solve these pairs of simultaneous equations.

a $4x + 5y = 16$
  $6x + 2y = 13$

b $3c + 8d = 7$
  $5c - 6d = 2$

c $6x + 5y = 9$
  $3y = 6 - 4x$

d $4g + 3h = -20$
  $3g - 2h = 2$

e $3x - 2y = 12$
  $5x - 9y = 20$

f $7w - 4z = 37$
  $2w + 3z = -6$

g $5x - 2y = 9$
  $2x = 8 + 3y$

h $\frac{1}{3}p + \frac{1}{5}q = 7$
  $\frac{3}{4}p + \frac{2}{3}q = 19$

10 Mandy and Zara are trying to solve the simultaneous equations:

  $4x + 3y = 5$
  $3x - 2y = -9$

Mandy says that you multiply the first equation by 2, multiply the second equation by 3 and add the new equations.
Zara disagrees: she says that you multiply the first equation by 3, multiply the second by 4 and subtract the new equations.

a Solve the equations using:

  i Mandy's method
  ii Zara's method.

b Who is correct? Give a reason for your answer.

11 a Solve the simultaneous equations $10x - 4y = 19$ and $4x - 3y = 9$.

  b On graph paper, draw axes from −5 to 5.
  Plot the graphs of $y = \frac{5}{2}x - \frac{19}{4}$ and $y = \frac{4}{3}x - 3$.
  What is their point of intersection?

  c What is the connection between your answers to parts **a** and **b**?

12 Get Real!
A taxi firm charges a fixed amount, £f, for each journey and then m pence a mile.
A 4-mile journey costs £5.40
A 7-mile journey costs £7.80
How much will a 9-mile journey cost?

13 Without drawing the graphs, find the coordinates of the point of intersection of the lines $5y = 6 - 3x$ and $3y = 5 - 2x$.

14 Daniel is confused!
He is trying to solve the simultaneous equations $2x + 2y = 5$ and $3y = 4 - 3x$.

  a Plot the graphs of $2x + 2y = 5$ and $3y = 4 - 3x$.

  b Why can't Daniel solve the equations simultaneously?

15 Get Real!
Three friends are downloading music from a web site.
Arminder pays £18.13 to download 3 albums and 4 single tracks.
Becky pays £13.93 to download 2 albums and 5 single tracks.
Charlotte wants to download 4 albums and 3 single tracks.
How much will Charlotte be charged?

16 Get Real!

The Pickard family, Mr and Mrs Pickard and their three children pay £38.50 to get into a wildlife park.
The Shabir family, Mr and Mrs Shabir, their grown-up daughter and four children pay £54.50 to get in.
How much are adult and child tickets for the park?

17 Sue is thinking of a fraction, $\frac{x}{y}$

If she adds 4 to the numerator and the denominator her fraction equals $\frac{3}{5}$

If she subtracts 6 from the numerator and the denominator, her fraction equals $\frac{1}{3}$
What is Sue's fraction?

18 ABC is an equilateral triangle.
Form two simultaneous equations and solve them to find a, b and the perimeter of the triangle.

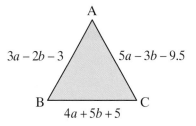

Explore

- ◎ Solve these two special 'matching' simultaneous equations:
  $$55x + 45y = 75$$
  $$45x + 55y = 25$$

- ◎ It is hard work like that, isn't it? Try this method:
  – Add the two equations and divide by the coefficient of x (100)
  – Subtract the two equations and divide by the coefficient of x (10)
  – Now you have two much simpler simultaneous equations to add and solve

- ◎ Make up some other complicated equations with matching x- and y-coefficients

- ◎ Try challenging your friends to solve them

Investigate further

Explore

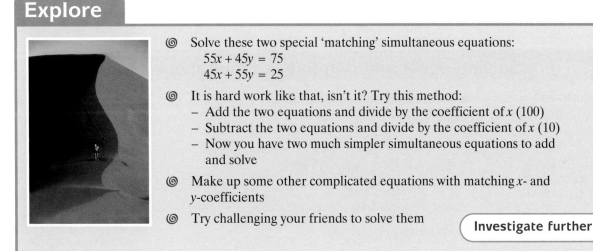

- ◎ So far you have solved a pair of simultaneous equations with two unknowns using two equations

- ◎ If you are trying to solve three simultaneous equations with three unknowns you need three equations

- ◎ If you are trying to solve four simultaneous equations with four unknowns you need four equations ... etc

- ◎ Here is a set of three simultaneous equations with three unknowns:

  $$a + b + c = 7$$
  $$a + b - c = 5$$
  $$a - b + c = 3$$

  HINT Try adding pairs of equations.

- ◎ Solve the simultaneous equations to find a, b and c

Investigate further

Explore

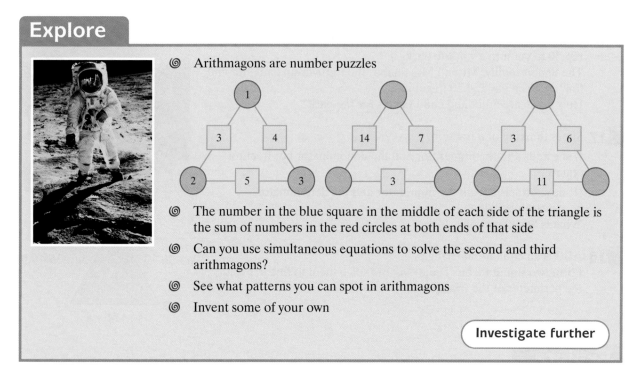

◎ Arithmagons are number puzzles

◎ The number in the blue square in the middle of each side of the triangle is the sum of numbers in the red circles at both ends of that side

◎ Can you use simultaneous equations to solve the second and third arithmagons?

◎ See what patterns you can spot in arithmagons

◎ Invent some of your own

Investigate further

Learn 4 Solving linear and non-linear simultaneous equations

Examples:

a Solve the simultaneous equations:

$y = 3x - 2$ — Linear

$y = x^2$ — Non-linear

Since $y = 3x - 2$ and $y = x^2$, then:

$$3x - 2 = x^2$$

and $x^2 - 3x + 2 = 0$ Subtracting $3x - 2$ from both sides

Factorising and solving:

$$(x - 2)(x - 1) = 0$$

$$(x - 2) = 0 \quad \text{or} \quad (x - 1) = 0$$

$$x = 2 \quad \text{or} \quad x = 1$$

Substituting into the *linear* equation:

when $x = 2$, $y = 3x - 2$
$y = 3 \times 2 - 2 = 4$

when $x = 1$, $y = 3x - 2$
$y = 3 \times 1 - 2 = 1$

So the solutions are $x = 1$ and $y = 1$ or $x = 2$ and $y = 4$.

An alternative approach would be to use $x = \dfrac{y + 2}{3}$

from $y = 3x - 2$ and substitute in $y = x^2$ ... but the first method is much easier

b Solve the simultaneous equations to find the points of intersection of their graphs.

$$x^2 + y^2 = 26$$
$$y = x - 6$$

$$x^2 + y^2 = 26 \longleftarrow \text{Non-linear}$$
$$y = x - 6 \longleftarrow \text{Linear}$$

Substituting for *y from the linear equation* into the non-linear equation:

$$x^2 + (x - 6)^2 = 26$$

Expanding and rearranging:

$$(x - 6)^2 = x^2 - 6x - 6x + 36$$
$$= x^2 - 12x + 36$$

$$x^2 + x^2 - 12x + 36 = 26$$
$$2x^2 - 12x + 10 = 0 \qquad \text{Divide through by 2}$$
$$x^2 - 6x + 5 = 0$$

Factorising and solving:

$$(x - 5)(x - 1) = 0$$
$$x - 5 = 0 \quad \text{or} \quad x - 1 = 0$$
$$x = 5 \quad \text{or} \qquad x = 1$$

To check your solutions substitute into the non-linear equation

$$(5)^2 + (-1)^2 = 26 \ \checkmark$$
$$(1)^2 + (-5)^2 = 26 \ \checkmark$$

Substituting into the *linear* equation:

when $x = 5$, $\quad y = 5 - 6 = -1$
when $x = 1$, $\quad y = 1 - 6 = -5$

So the solutions are $x = 5$ and $y = -1$ or $x = 1$ and $y = -5$.
The coordinates of the points of intersection are $(5, -1)$ and $(1, -5)$.

You have found the two points of intersection of the circle and the straight line

Apply 4

1 Solve the following pairs of simultaneous equations.

a $y = 7x + 2$
$\quad y = 4x^2$

b $\quad y = 2x^2$
$\quad y + 5x = 3$

c $\quad y = 3x^2$
$\quad y + 14x = 5$

2 By solving the following pairs of equations simultaneously, find the coordinates of the points of intersection of their graphs.

a $xy = 8$
$\quad y = x + 2$

b $2y = x + 5$
$\quad xy = 12$

c $xy + 6 = 0$
$\quad 3y + 2x = 0$

225

3 Get Real!

Ros is making a rectangular run for her pet rabbit.
She has 5 metres of fencing to make the run.
She wants the run to have an area of 1.5 square metres.
Set up two simultaneous equations in x and y, the length and width of the run.
What are the dimensions of Ros' rabbit run?

4 Get Real!

Daksha has bought two printer ink cartridges from a web site.
He pushes the wrong button on his calculator when he's working out the total cost and multiplies the prices instead of adding.
He gets the result £108.
The correct total is £21.
How much is each ink cartridge?

5 Solve the following pairs of simultaneous equations.

a $x^2 + y^2 = 16$
$\quad\quad y = x - 4$

d $x^2 + y^2 = 20$
$\quad\quad y = 2x - 6$

b $x^2 + y^2 = 169$
$\quad\quad x + y = 7$

e $x^2 + y^2 = 17$
$\quad\quad 3y + x = 11$

c $x^2 + y^2 = 2$
$\quad\quad x + y = 0$

f $x^2 + y^2 = 25$
$\quad\quad y = 2x - 2$

6 Solve the following pairs of simultaneous equations.

a $y = x^2 + x + 1$
$\quad y = 6x + 15$

b $y = 2x^2 - 7x + 6$
$\quad y = 2x + 11$

c $\quad y = 6x^2 + 5x$
$\quad 10 = 6x + y$

7 Sam is solving the simultaneous equations $y = 2x^2 - 3x - 3$ and $x + y = 1$.
He substitutes $x = 1 - y$ into the quadratic equation and gets
$y = 2(1 - y)^2 - 3(1 - y) - 3$.
Complete his method to find y and then x.
Can you suggest a simpler method that Sam could use?

8 I am thinking of two integers.
The sum of their squares is 80.
The first number is four more than the second number.
What could my two numbers be?

9 a Factorise $4x^2 - y^2$.

b Hence, or otherwise, solve the simultaneous equations:

$4x^2 - y^2 = 32$
$\quad 2x + y = 4$

Explore

◎ Solve these pairs of simultaneous equations

a $y = x^2 + 2x + 2$
$y = 3x + 4$

b $y = x^2 + 2x + 2$
$y = 4x + 1$

◎ On graph paper, draw an x-axis from -4 to 2 and a y-axis from -2 to 12

◎ Plot the curve $y = x^2 + 2x + 2$, and the lines $y = 3x + 4$, and $y = 4x + 1$

◎ How does the graph help explain what happened with sets of equations **a** and **b**?

◎ Try to solve the simultaneous equations $y = x^2 + 2x + 2$ and $y = x + 1$

Investigate further

Explore

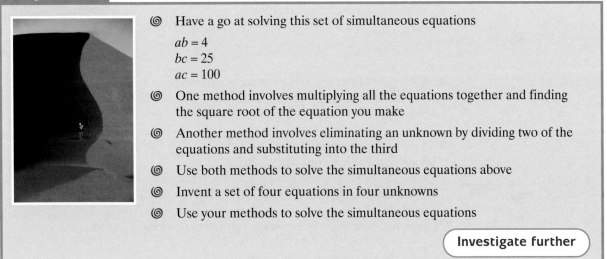

◎ Have a go at solving this set of simultaneous equations

$ab = 4$
$bc = 25$
$ac = 100$

◎ One method involves multiplying all the equations together and finding the square root of the equation you make

◎ Another method involves eliminating an unknown by dividing two of the equations and substituting into the third

◎ Use both methods to solve the simultaneous equations above

◎ Invent a set of four equations in four unknowns

◎ Use your methods to solve the simultaneous equations

Investigate further

Inequalities and simultaneous equations

ASSESS

The following exercise tests your understanding of this chapter, with the questions appearing in order of increasing difficulty.

1 Show each of the following inequalities on a number line.

a $x > -2$

c $4 < x \leqslant 9$

b $x < 4$

d $-3 \leqslant x \leqslant 7$

2 List all the **integer** solutions to the inequalities:

a $2 < 3x < 14$

c $15 \geqslant 3z \geqslant -4$

b $-7 \leqslant 4a \leqslant 5$

3 Solve these inequalities:

a $3a < 12$ **c** $4g - 2 > -3$

b $4b \geqslant -20$ **d** $3x - 2 > 4x + 1$

4 Shade the region defined by the three inequalities $2y < 3x + 11$, $3y + 4x \leqslant 12$ and $x \leqslant 4$.

5 Solve the simultaneous equations:

$$2a + 3b = 19$$
$$3a - b = 1$$

6 Batman and Robin are b and r years old respectively.
The sum of their ages is 44.
Batman is 8 years older than Robin.
Write down equations in b and r and solve them to find the respective ages of the dynamic duo!

7 Solve the simultaneous equations:

$$3m + 4n = 37$$
$$2m - 5n = -6$$

8 The ages of Florence and Zebedee are in the ratio $2 : 3$.
In 4 years' time their ages be in the ratio $3 : 4$.
How old are Florence and Zebedee now?

9 A curve has the equation $xy = 16$.
A straight line has equation $y = 10 - x$.
Solve the two equations simultaneously to find the coordinates of the points where the straight line intersects the curve.

Try some real past exam questions to test your knowledge:

10 Solve the simultaneous equations:

$$y = x + 7$$
$$x^2 + y^2 = 25$$

You **must** show your working.
Do **not** use trial and improvement.

Spec B, Module 5, Higher Paper 1, June 03

11 A straight line has the equation $y = 2x - 3$
A curve has the equation $y^2 = 8x - 16$

Solve these simultaneous equations to find any points of intersection of the line and the curve.

You **must** show all your working.
Do **not** use trial and improvement.

Spec A, Higher Paper 1, June 05

16 Trigonometry

B **Examiners would normally expect students who get a B grade to be able to:**

Use sine, cosine and tangent to calculate a side in a right-angled triangle

Use sine, cosine and tangent to calculate an angle in a right-angled triangle

A **Examiners would normally expect students who get an A grade also to be able to:**

Sketch and draw trigonometric graphs

Use the sine rule and the cosine rule to find the missing sides and missing angles in any triangle

Use the formula for the area of a non right-angled triangle

A* **Examiners would normally expect students who get an A* grade also to be able to:**

Use trigonometry to find sides and angles in three dimensions

Understand the graphs of trigonometric functions for angles of any size

Find the angle between a line and a plane

What you should already know ...

- The sum of angles in a triangle and angles round a point
- Symmetry and angle properties of isosceles triangles
- Measure angle bearings
- Ratio properties of similar triangles
- Pythagoras' theorem

Trigonometry – the branch of mathematics that deals with the relationship between the lengths of sides and sizes of angles in triangles

Opposite side – in a right-angled triangle, the side opposite the known angle

Opposite

Adjacent side – in a right-angled triangle, the shorter side adjacent to the known angle

Adjacent

Hypotenuse – the longest side of a right-angled triangle, opposite the right angle

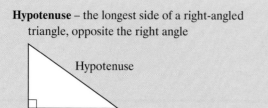

Hypotenuse

Sine (abbreviation sin) – in a right-angled triangle, the ratio of the length of the opposite side to the length of the hypotenuse

$$\sin x = \frac{opposite}{hypotenuse}$$

Cosine (abbreviation cos) – in a right-angled triangle, the ratio of the length of the adjacent side to the length of the hypotenuse

$$\cos x = \frac{adjacent}{hypotenuse}$$

Tangent (abbreviation tan) – in a right-angled triangle, the ratio of the length of the opposite side to the length of the adjacent side

$$\tan x = \frac{opposite}{adjacent}$$

Angle of elevation – the angle above the horizontal between a line and the horizontal. Example: in the diagram, the angle b is the angle of elevation of point A on top of the cliff from point B

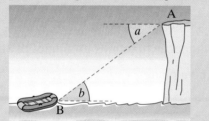

Angle of depression – the angle below the horizontal between a line and the horizontal. Example: in the diagram, the angle a is the angle of depression of point B from point A

Isosceles triangle – a triangle with 2 equal sides and 2 equal angles; the equal angles are called **base angles**

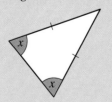

Bearing – an angle measured clockwise from North; all bearings should be written as three figure numbers, for example, 125° or 045°

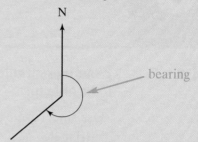

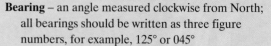

bearing

Trigonometric functions – the sine, cosine and tangent functions, $y = \sin x$, $y = \cos x$ and $y = \tan x$, are the three most common of these

The values of the sine and the cosine functions repeat every 360° and the values of the tangent function repeat every 180°, so the functions are described as **periodical**

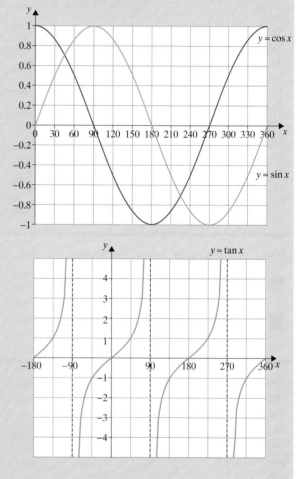

Learn 1   Using sine, cosine and tangent to calculate a side

Examples: **a** Calculate the length of the side AB.

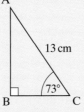

Copy and label the diagram.

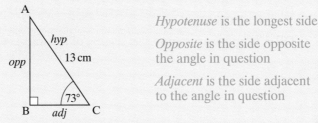

Hypotenuse is the longest side

Opposite is the side opposite the angle in question

Adjacent is the side adjacent to the angle in question

Identify the known side (*hyp*) and the required side (*opp*) to establish whether to use sin, cos or tan

Here $\sin 73° = \dfrac{opp}{hyp}$ $\cos 73° = \dfrac{adj}{hyp}$ $\tan 73° = \dfrac{opp}{adj}$

You know *hyp* and want to find *opp*, so you use sin

$\sin 73° = \dfrac{opp}{hyp}$ ⟵ ———————— Start with the formula

$0.95630475\ldots = \dfrac{x}{13}$ ⟵ AB ——— Substitute known facts

$0.95630475\ldots \times 13 = x$ ⟵ AB —— Solve the equation by multiplying both sides by 13

$AB = 12.4$ cm (to 3 s.f.) ⟵ ——— Choose an appropriate degree of accuracy – 3 s.f. are usually sufficient

b Calculate the length of the side DE.
Give your answer to an appropriate degree of accuracy.

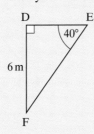

Copy and label the diagram.

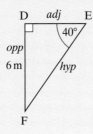

Identify the known side (*opp*) and required side (*adj*) to establish whether to use sin, cos or tan

Here $\sin 40° = \dfrac{opp}{hyp}$ $\cos 40° = \dfrac{adj}{hyp}$ $\tan 40° = \dfrac{opp}{adj}$

You know *opp* and want to find *adj*, so you use tan

$\tan 40° = \dfrac{opp}{adj}$ ←——————— Start with the formula

$0.83909963 ... = \dfrac{6}{DE}$ ←——————— Substitute known facts

$DE \times 0.83909963 ... = 6$ ←——————— Multiply both sides by DE

$DE = \dfrac{6}{0.83909963 ...}$ ←——————— Divide both sides by 0.83909963 ...

$DE = 7.2$ m (to an appropriate degree of accuracy) ←—— As original data to nearest whole number

Apply 1

Apart from question 8 this is a calculator exercise.

1 Use your calculator to find these.
Give your answers to 3 significant figures.

    a $\sin 70°$ **c** $\tan 41°$ **e** $\cos 34.5°$

    b $\cos 30°$ **d** $\tan 71°$ **f** $\sin 53.1°$

2 Copy each diagram and label it. Calculate the length marked *x*; giving your answers to 3 s.f.

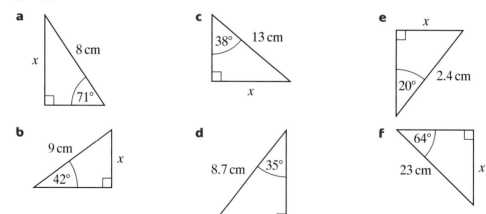

3 Copy the diagram and label it in each case. Calculate the side labelled x, giving your answers to 1 d.p.

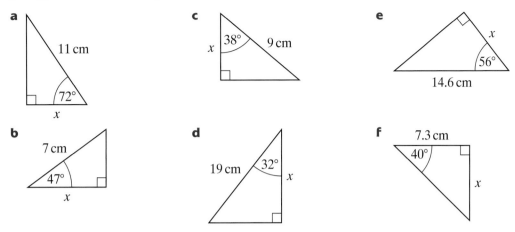

a 11 cm, 72°, x

c 38°, 9 cm, x

e x, 56°, 14.6 cm

b 7 cm, 47°, x

d 19 cm, 32°, x

f 7.3 cm, 40°, x

4 Get Real!
From take-off, an aeroplane climbs at an angle of 20°.
When the aeroplane has flown 10 km, what height has it reached?

5 This right-angled triangle has a hypotenuse of 5.4 cm.

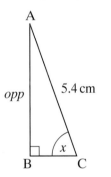

A, opp, 5.4 cm, x, B, C

Choose five possible values of x, and for each one calculate the length of the opposite side.

6 Alison draws an equilateral triangle with sides of 8 cm.
She calculates the area, using the formula $A = \frac{1}{2}bh$
She says the area is $\frac{1}{2} \times 8 \times 8 = 32$ cm^2.

a Explain why Alison is wrong.

b Calculate the area of the triangle.

7 ABC and ACD are right-angled triangles.
AB = 13 cm.

a Calculate AC.

b Calculate DC.

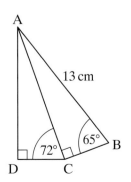

A, 13 cm, 72°, 65°, B, D, C

8 ABC is an isosceles triangle.

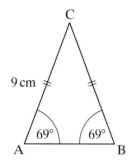

Given that sin 69° = 0.93 (2 d.p.), cos 69° = 0.36 (2 d.p.) and
tan 69° = 2.61 (2 d.p.), calculate the base, AB.
Give your answer to an appropriate degree of accuracy.

9 Get Real!

A sports pavilion is designed with a sloping roof.

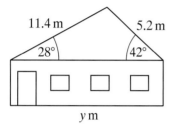

Using the information given on the diagram, calculate the length of the
pavilion, marked *y* m.

10 Find the length of the hypotenuse in each of these triangles.

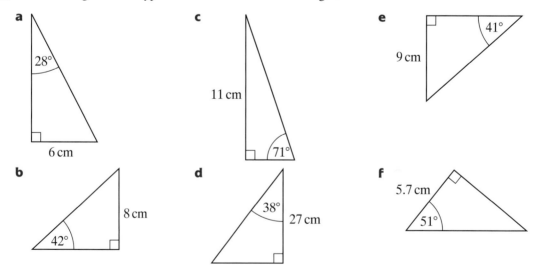

Explore

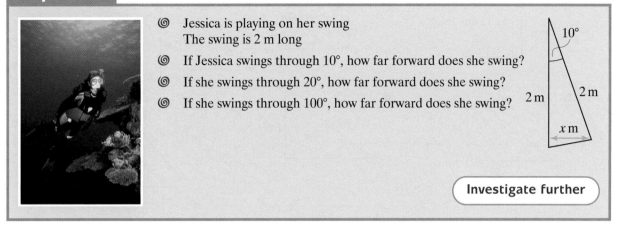

◎ Jessica is playing on her swing
The swing is 2 m long

◎ If Jessica swings through 10°, how far forward does she swing?

◎ If she swings through 20°, how far forward does she swing?

◎ If she swings through 100°, how far forward does she swing?

Investigate further

Learn 2   Using sine, cosine and tangent to calculate an angle

Example: Find the size of angle *x* in this diagram.

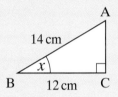

Copy and label the diagram.

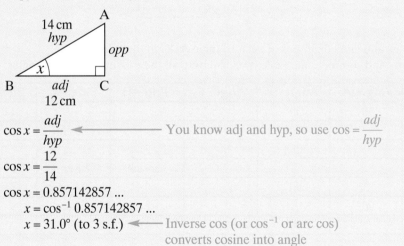

$$\cos x = \frac{adj}{hyp}$$ ⟵ You know adj and hyp, so use $\cos = \frac{adj}{hyp}$

$$\cos x = \frac{12}{14}$$

$$\cos x = 0.857142857 \ldots$$

$$x = \cos^{-1} 0.857142857 \ldots$$

$$x = 31.0° \text{ (to 3 s.f.)}$$ ⟵ Inverse cos (or $\cos^{-1}$ or arc cos) converts cosine into angle

Apply 2

Apart from question 5 this is a calculator exercise.

1 Write the angle to one decimal place, whose sine is:

   a 0.8   **c** 0.45   **e** 0.02

   b 0.3   **d** 0.2345   **f** 0.956

2 Calculate the size of the angles labelled x in these diagrams.

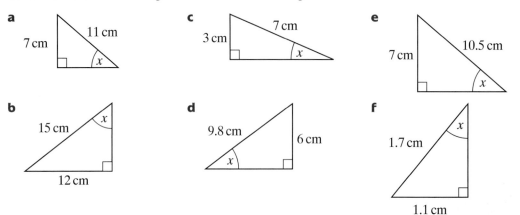

a 7 cm, 11 cm, x

c 7 cm, 3 cm, x

e 7 cm, 10.5 cm, x

b 15 cm, 12 cm, x

d 9.8 cm, 6 cm, x

f 1.7 cm, 1.1 cm, x

3 Calculate the size of the angles labelled x in these diagrams.

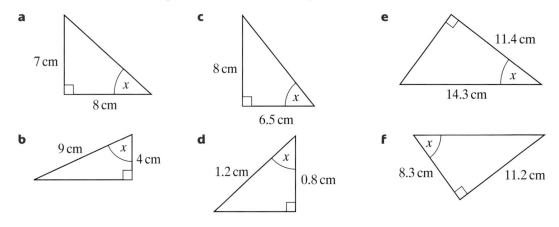

a 7 cm, 8 cm, x

c 8 cm, 6.5 cm, x

e 11.4 cm, 14.3 cm, x

b 9 cm, 4 cm, x

d 1.2 cm, 0.8 cm, x

f 8.3 cm, 11.2 cm, x

4 Draw and label diagrams of each triangle ABC in the table.
Copy and complete the table by calculating all the missing sides and angles.

	Angle A	Angle B	Angle C	Side AB	Side AC	Side BC
a	90°	30°				11 cm
b		90°	43°		9 cm	
c		28°	90°	13.5 cm		
d			90°	11.4 cm	7.8 cm	
e		90°			3.4 cm	2.9 cm
f	90°			3.2 cm	4.5 cm	
g		90°	28°	7.4 cm		

5 Terri says that sin 45° is the same as cos 45°. Is she right?
Give a reason for your answer.

6 **Get Real!**

Maythorpe is 9 km due west of Hopton.
Hopton is due north of Colton.
The distance form Maythorpe to Colton is 14.7 km.
Calculate the bearing of Colton from Maythorpe.

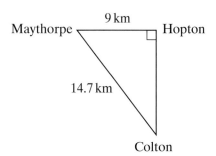

Maythorpe — 9 km — Hopton
14.7 km
Colton

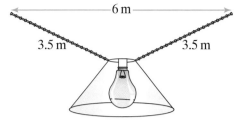

7 Get Real!

A light fitting hangs from the two chains 3.5 m long.
The ends of the chains are 6 m apart.
What is the angle between the two chains?

Explore

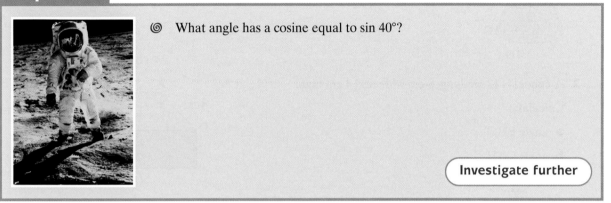

◎ What angle has a cosine equal to sin 40°?

Investigate further

Learn 3 Trigonometry in three dimensions

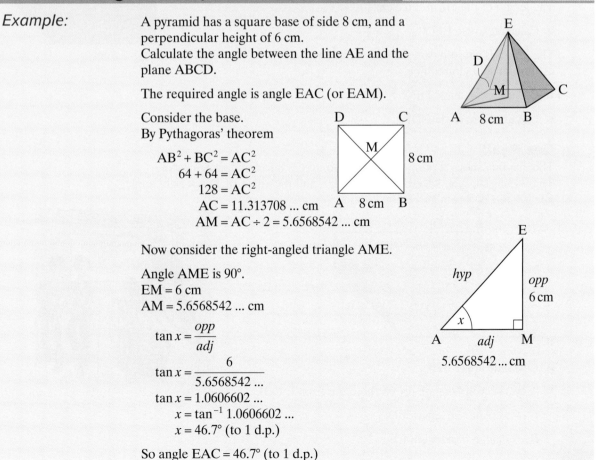

Example:

A pyramid has a square base of side 8 cm, and a
perpendicular height of 6 cm.
Calculate the angle between the line AE and the
plane ABCD.

The required angle is angle EAC (or EAM).

Consider the base.
By Pythagoras' theorem

$$AB^2 + BC^2 = AC^2$$
$$64 + 64 = AC^2$$
$$128 = AC^2$$
$$AC = 11.313708 \dots \text{ cm}$$
$$AM = AC \div 2 = 5.6568542 \dots \text{ cm}$$

Now consider the right-angled triangle AME.

Angle AME is 90°.
EM = 6 cm
AM = 5.6568542 ... cm

$$\tan x = \frac{opp}{adj}$$
$$\tan x = \frac{6}{5.6568542 \dots}$$
$$\tan x = 1.0606602 \dots$$
$$x = \tan^{-1} 1.0606602 \dots$$
$$x = 46.7° \text{ (to 1 d.p.)}$$

So angle EAC = 46.7° (to 1 d.p.)

Apply 3

1 A cone has a height of 9 cm and a circular base of radius 4 cm.
Calculate the angle ACB.

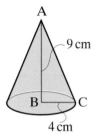

2 A cuboid is 12 cm long, 8 cm wide and 4 cm high.

Calculate:

a angle BHG

b angle CAH.

3 The diagram shows an eraser in the shape of a prism.
The front, DEFG, is a right-angled trapezium.
AB is 4 cm long and is parallel to GF which is 7.6 cm long.
DG = 3 cm.
The length of the prism is 6.2 cm.

a Calculate angle EFG.

The eraser is cut into two, through the plane ACFD,
as shown by the dotted line.

b Calculate the angle DFG.

c Calculate the angle CDF.

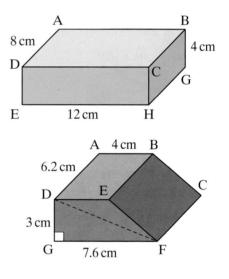

4 Get Real!

The diagram shows a ramp for wheelchair access into a shopping centre.
The sloping surface, ABDE, is rectangular, and the angle BCD is a right angle.
ED is 6 m, CD is 4 m and BC is 1.4 m.

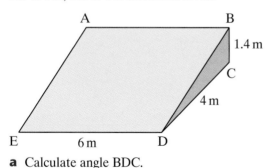

a Calculate angle BDC.

b A cyclist cycles diagonally up the ramp from E to B.

i Calculate the length BD.

ii Calculate the length EB.

iii Calculate the angle BEC.

5 A pencil rests in a hollow cube, which measures 6 cm × 6 cm × 6 cm. Calculate the angle the pencil makes with the horizontal.

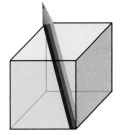

6 ABCDE is a square-based pyramid. BC = 9 cm and AC = 11 cm.

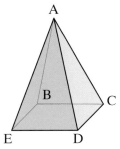

Calculate:

a angle BAC

b angle BAD.

7 A tent is in the shape of an isosceles triangular prism. DE = EF = 1.8 m. The angle EDF = 64°.

a Calculate the height of the tent.

Two guy ropes, EX and EY, are each 2.3 m long.

b Calculate the angle EY makes with the horizontal.

The ends X and Y are pegged so that the distance XY is equal to FD.

c Calculate the angle YEX.

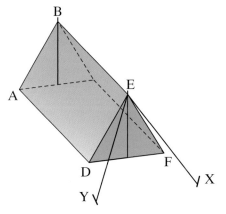

8 Calculate the angle between the diagonal of a cube and an adjacent edge of a cube.

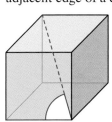

Explore

- On a sheet of graph paper, draw axes going from −10 to 10, using a scale of 1 cm to 1 unit
- Draw a circle, centre *O*, with a radius of 10 cm
- Draw a radius in the first quadrant
- What are the coordinates of the point where the radius meets the circle?

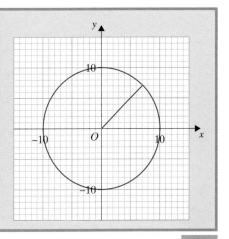

Investigate further

Learn 4 Going beyond 90°

Examples: Solve the equations:

a $\sin x = 0.47$ **b** $3\cos x = -0.69$

for values of x between 0° and 360°.

The symmetry of the graph tells you that $\sin x = \sin(180 - x)$ and $\sin x = -\sin(180 + x)$

a

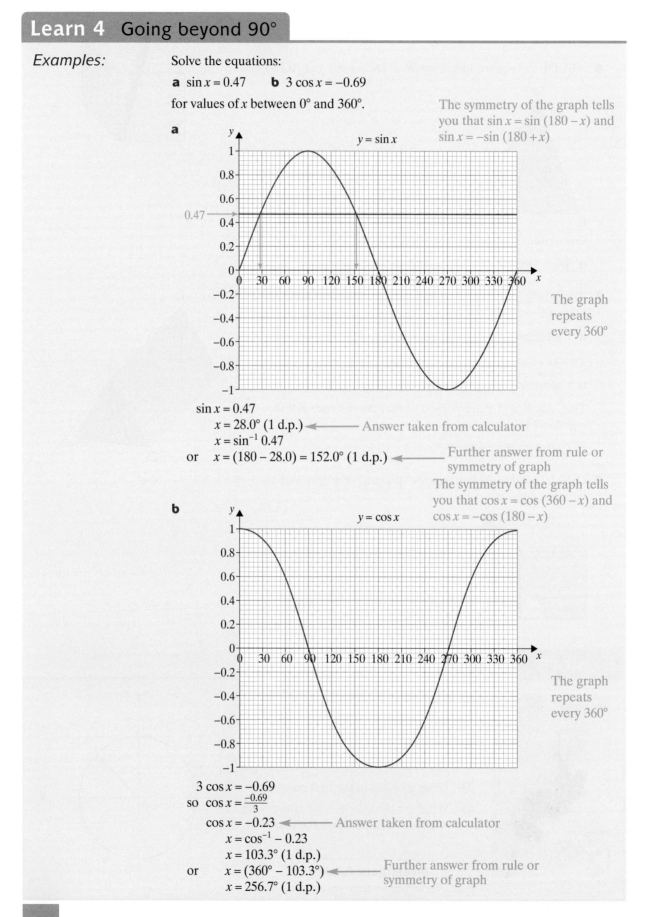

$y = \sin x$

The graph repeats every 360°

$\sin x = 0.47$

$\qquad x = 28.0°$ (1 d.p.) $\longleftarrow$ Answer taken from calculator

$\qquad x = \sin^{-1} 0.47$

or $\quad x = (180 - 28.0) = 152.0°$ (1 d.p.) $\longleftarrow$ Further answer from rule or symmetry of graph

The symmetry of the graph tells you that $\cos x = \cos(360 - x)$ and $\cos x = -\cos(180 - x)$

b

$y = \cos x$

The graph repeats every 360°

$3\cos x = -0.69$

so $\cos x = \dfrac{-0.69}{3}$

$\qquad \cos x = -0.23$ $\longleftarrow$ Answer taken from calculator

$\qquad x = \cos^{-1} - 0.23$

$\qquad x = 103.3°$ (1 d.p.)

or $\quad x = (360° - 103.3°)$ $\longleftarrow$ Further answer from rule or symmetry of graph

$\qquad x = 256.7°$ (1 d.p.)

Apply 4

1 Sketch the graph of $y = \sin x$ for values of x from $0°$ to $720°$.

2 Sketch the graph of $y = \cos x$ for values of x from $0°$ to $720°$.

3 Which angles between $0°$ and $360°$ have the same sine as:

    a $30°$ **d** $200°$ **g** $199°$

    b $78°$ **e** $325°$ **h** $522°$

    c $144°$ **f** $400°$

4 Which angles between $0°$ and $360°$ have the same cosine as:

    a $40°$ **d** $111°$ **g** $222°$

    b $69°$ **e** $312°$ **h** $555°$

    c $99°$ **f** $410°$

5 Phil says that $\sin 100° = \sin 260°$
Jill says $\sin 100° = -\sin 260°$
Will says $\cos 100° = \cos 260°$
Who is right? Give a reason for your answer.

6 Solve these equations, giving all the answers in the range $0°$ to $360°$.
Give your answers to an appropriate degree of accuracy.

    a $\sin x = 0.7$ **c** $\sin x = 0.5$ **e** $\cos x = 0.85$

    b $\cos x = 0.23$ **d** $\cos x = -0.23$ **f** $\sin x = -0.76$

7 Solve these equations, giving all answers in the range $0°$ to $360°$.
Give your answers to an appropriate degree of accuracy.

    a $4 \sin x = 3$ **c** $5 \cos x = -2$ **e** $7 \sin x = -4$

    b $2 \cos x = 1$ **d** $8 \cos x = -3$ **f** $9 \sin x = -5$

8 Get Real!

A helicopter circles over a village. It starts facing due east, and travels anticlockwise in a circle of radius of 5 km.

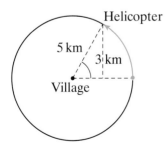

    a What angle has it turned through when it is 3 km north of the village?

    b What angle will it have turned through when it is next 3 km north of the village?

    c What angle has it turned through when it is 3 km south of the village?

    d What angle will it have turned through when it is next 3 km south of the village?

9 Plot graphs of $y = \sin x$ and $y = \cos x$ for $0° \leqslant x \leqslant 360°$ on the same diagram.

   a Use your diagram to solve the equation $\sin x = \cos x$ for $0° \leqslant x \leqslant 360°$

   b Use the answers from part **a** to calculate the solutions to the equation $\sin x = \cos x$ for $360° \leqslant x \leqslant 720°$

10 Use the graphs from question **9** for this question.
Which of these statements are correct?

   a $\sin 45° = \cos 45°$ **e** $\cos 120° = -\cos 60°$

   b $\sin 45° = \sin 135°$ **f** $\sin 60° = \sin 420°$

   c $\cos 45° = \cos 135°$ **g** $\cos 60° = \cos 420°$

   d $\cos 60° = \sin 30°$

11 Which of these statements are correct?

   a $\sin x = \sin (180° - x)$ **d** $\sin x = -\sin (-x)$

   b $\cos x = \cos (180° - x)$ **e** $\cos x = -\cos (-x)$

   c $\sin x = \cos (90° - x)$

   In each case, give an example to illustrate whether the statement is true or false.

12 a Copy and complete this table of values for $y = \tan x$.

x	0	30	60	75	90	105	120	150	180	210	240	255	270	285	300	330	360
y	0				—				0				—				0

   b Draw the x-axis for values from $0°$ to $360°$ and the y-axis for values from -4 to 4.
   Plot the points and join them with smooth curves.

   c After how many degrees does the graph of the tangent function repeat itself?

   d Use the graph to find approximate solutions to:

       i $\tan x = 2$ **iii** $\tan x = -2$

       ii $\tan x = \frac{1}{2}$ **iv** $\tan x = -\frac{1}{2}$

   e Use your calculator to find the solutions to one decimal place.

13 Use the graph from question **12** for this question.
Which of these statements are correct?

   a $\tan 45° = \tan 225°$ **d** $\tan 0° = \cos 0°$

   b $\tan 30° = -\tan (-30)°$ **e** $\tan 60° = 2 \tan 30°$

   c $\tan 0° = \sin 0°$

14 a Liz says that if you double an angle, you double the sine.
   Give an example to show that Liz is not correct.

   b Phil says that, if two angles add up to $90°$, the sine of one angle is equal to the cosine of the other. Is Phil correct? Give a reason for your answer.

   c Jake says, 'As the angle gets bigger, the sine gets bigger and the cosine gets smaller.' Is Jake correct? Give a reason for your answer.

15 Sketch a graph of $y = \sin x$ on axes scaled from 0° to 360° for x and from −5 to 4 for y. On the diagram, sketch graphs of:

  a $y = 2 \sin x$

  b $y = 3 + \sin x$

  c $y = 3 \sin x - 2$

16 Use the fact that sin 30 = 0.5 to solve these equations for $0° \leqslant x \leqslant 360°$

  a $\sin 2x = 0.5$

  b $2 \sin x + 1 = 0$

  c $3 \sin 3x + 3 = 4.5$

17 Get Real!

The diagram shows the depth of water above the low water marker in a harbour during a 24-hour period. The blue line shows the actual values. The red line shows the cosine function $y = 1.4\cos (x + 1)\frac{90}{\pi} + 2.2$, which gives a close approximation to the real data.

Depth of water above low water level

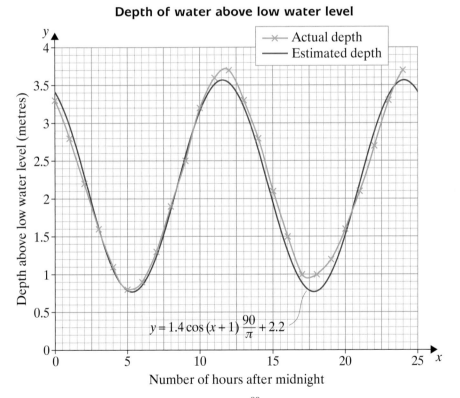

a Show that the function $y = 1.4 \cos (x + 1)\frac{90}{\pi} + 2.2$ gives an estimated depth of approximately 3.43 metres at the start of the 24-hour period.

b i At what time does the cosine function give the least accurate estimate?

  ii Find the percentage error in the estimate compared with the actual depth at this time.

Explore

- ◎ On a piece of graph paper, draw x- and y-axes labelled from –1 to 1; make the scale as big as possible and the same on each axis
- ◎ Choose an angle (any size, positive or negative); find the cosine and sine of the angle and mark a point on your graph, using the cosine of the angle as the x-coordinate and the sine as the y-coordinate

Investigate further

Explore

- ◎ Copy and complete this table for tan x to 2 decimal places for x-values from 0 to 90°

$x°$	0	10	20	30	40	50	60	70	80	90
tan $x°$	0				0.84					

- ◎ Something strange happens at 90°; to investigate, copy and complete this table for x-values from 81° to 99°

$x°$	81	83	85	87	89	91	93	95	97	99
tan $x°$	0			19.08						

- ◎ What happens close to 90°?
  You may want to try other values very close to 90° – just above and just below

Investigate further

Explore

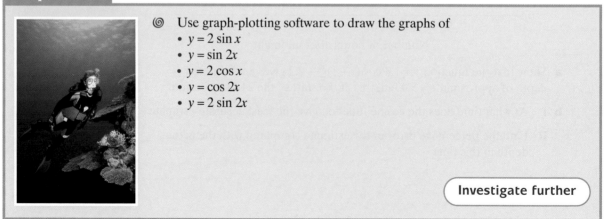

- ◎ Use graph-plotting software to draw the graphs of
  - $y = 2 \sin x$
  - $y = \sin 2x$
  - $y = 2 \cos x$
  - $y = \cos 2x$
  - $y = 2 \sin 2x$

Investigate further

Learn 5 The sine rule

Examples:

a Calculate angle A in the diagram below.

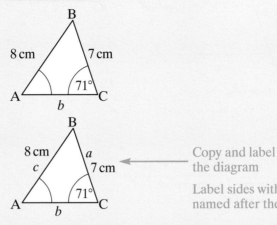

Copy and label the diagram

Label sides with lower case letters – named after the angle opposite them

The sine rule says

$$\frac{\sin A}{a} = \frac{\sin B}{b} = \frac{\sin C}{c}$$

or $\quad \dfrac{a}{\sin A} = \dfrac{b}{\sin B} = \dfrac{c}{\sin C}$

You need A; you know: a, c and C.

$$\frac{\sin A}{a} = \frac{\sin C}{c}$$

Select appropriate formula, with unknown in numerator

$$\frac{\sin A}{7} = \frac{\sin 71}{8}$$

$$\sin A = \frac{7 \times \sin 71°}{8}$$

$\sin A = 0.82732875 \ldots$

$A = \sin^{-1} 0.82732875 \ldots$

$A = 55.8°$ (to 1 d.p.)

Don't forget that $\sin x = 0.82732875 \ldots$ has two possible answers; $A = 55.8°$ or $A = 124.2°$. Use a diagram to decide which answer applies – or whether both answers could be right. Remember, the longest side is opposite the largest angle. In this case the angle $A = 124.2°$ is not possible as the angles can only all add up to $180° \ldots$ but it is always best to check

b Calculate the length of AB in the diagram below.

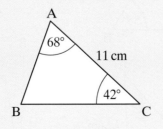

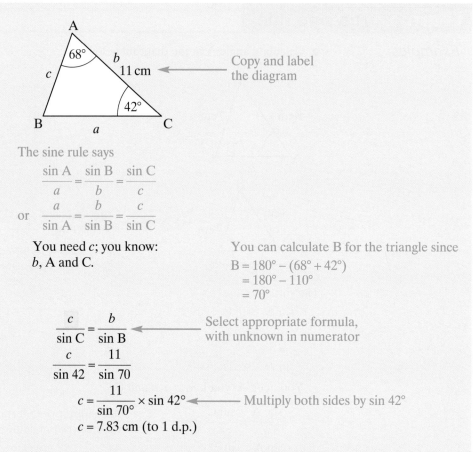

The sine rule says

$$\frac{\sin A}{a} = \frac{\sin B}{b} = \frac{\sin C}{c}$$

or $\dfrac{a}{\sin A} = \dfrac{b}{\sin B} = \dfrac{c}{\sin C}$

You need c; you know: b, A and C.

You can calculate B for the triangle since
$$B = 180° - (68° + 42°)$$
$$= 180° - 110°$$
$$= 70°$$

$$\frac{c}{\sin C} = \frac{b}{\sin B}$$

Select appropriate formula, with unknown in numerator

$$\frac{c}{\sin 42} = \frac{11}{\sin 70}$$

$$c = \frac{11}{\sin 70°} \times \sin 42°$$

Multiply both sides by sin 42°

$$c = 7.83 \text{ cm (to 1 d.p.)}$$

Remember the sine rule works in ANY triangle. It is used when the known facts and the required facts make up two angles and two sides.

Apply 5

1 Calculate the sides marked x in these diagrams.

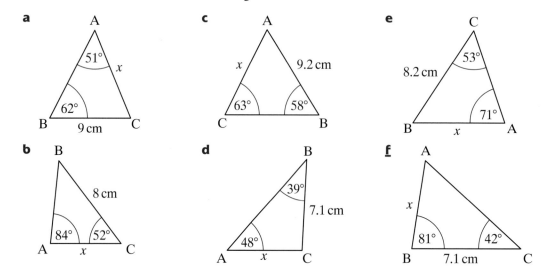

2 Calculate the angles marked x in these diagrams.

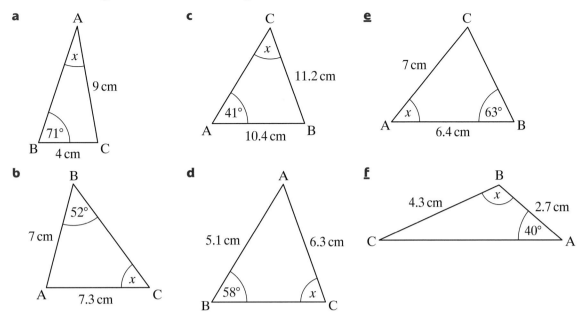

a A, x, 9 cm, 71°, B, 4 cm, C

c C, x, 11.2 cm, 41°, A, 10.4 cm, B

e C, 7 cm, x, 6.4 cm, 63°, A, B

b B, 52°, 7 cm, x, A, 7.3 cm, C

d A, 5.1 cm, 6.3 cm, 58°, x, B, C

f B, 4.3 cm, x, 2.7 cm, 40°, C, A

3 For each part, draw and label a diagram, and use the sine rule to find the side or angle (marked '?').

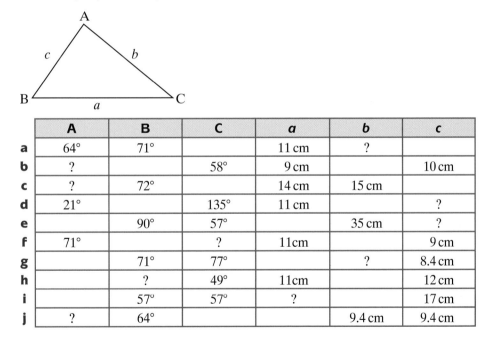

	A	B	C	a	b	c
a	64°	71°		11 cm	?	
b	?		58°	9 cm		10 cm
c	?	72°		14 cm	15 cm	
d	21°		135°	11 cm		?
e		90°	57°		35 cm	?
f	71°		?	11cm		9 cm
g		71°	77°		?	8.4 cm
h		?	49°	11cm		12 cm
i		57°	57°	?		17 cm
j	?	64°			9.4 cm	9.4 cm

4 Get Real!

Two coastguards see the same boat.

Coastguard B is 2.4 km due south of A.

The boat is on a bearing of 121° from coastguard A, and 073° from coastguard B.

Calculate the distance from each coastguard to the boat.

5 Kirsty is trying to answer this question:
'Triangle ABC has AB = 7 cm, AC = 10.8 cm, and angle ACB = 38°.
Calculate angle ABC.'
She gets an answer of 71.8°
She tries to construct the triangle using a ruler and compasses.
When she measures angle ABC to check her answer she finds that it is about 108°.
Which is the correct answer?

6 a In triangle ABC, angle A = 58°. BC = 9 cm and AB = 10 cm.
Find two possible sizes for angle C, and check your answers with accurate constructions.

b Triangle XYZ has XY = 7.2 cm and YZ = 8.4 cm.
If sin X = 0.74:

i what are the two possible sizes of angle X

ii what is the size of angle Z in each case?

Explore

◎ In triangle ABC, angle A = 55° and BC = 7 cm

◎ Can AB = 6 cm? If so, what is the size of angle C?

◎ Can AB = 8 cm? If so, what is the size of angle C?

◎ Can AB = 9 cm? If so, what is the size of angle C?

Investigate further

Explore

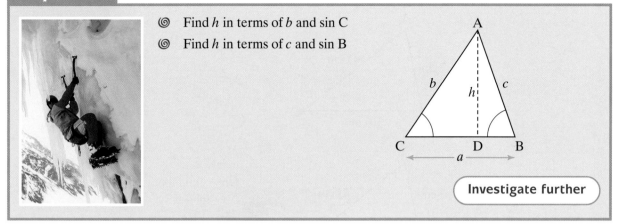

◎ Find h in terms of b and sin C

◎ Find h in terms of c and sin B

Investigate further

Learn 6   The cosine rule

Examples: **a** Calculate the length of AB in this triangle.

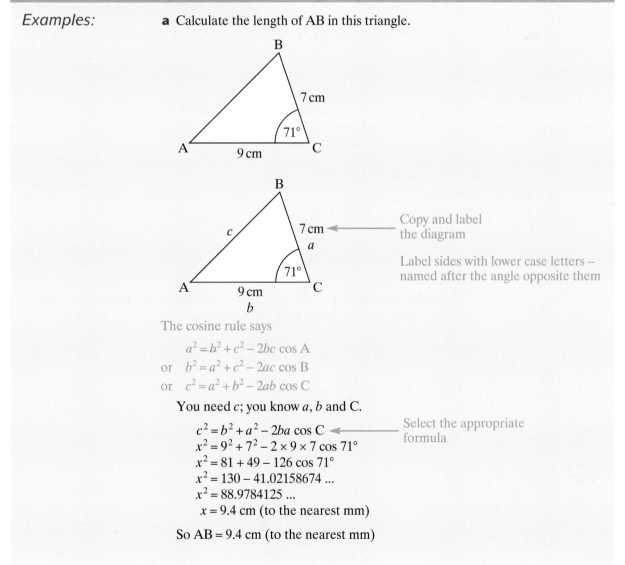

Copy and label the diagram

Label sides with lower case letters – named after the angle opposite them

The cosine rule says

$$a^2 = b^2 + c^2 - 2bc \cos A$$
or $$b^2 = a^2 + c^2 - 2ac \cos B$$
or $$c^2 = a^2 + b^2 - 2ab \cos C$$

You need c; you know a, b and C.

$$c^2 = b^2 + a^2 - 2ba \cos C$$
$$x^2 = 9^2 + 7^2 - 2 \times 9 \times 7 \cos 71°$$
$$x^2 = 81 + 49 - 126 \cos 71°$$
$$x^2 = 130 - 41.02158674 \ldots$$
$$x^2 = 88.9784125 \ldots$$
$$x = 9.4 \text{ cm (to the nearest mm)}$$

Select the appropriate formula

So AB = 9.4 cm (to the nearest mm)

b Calculate the size of angle B in this triangle.

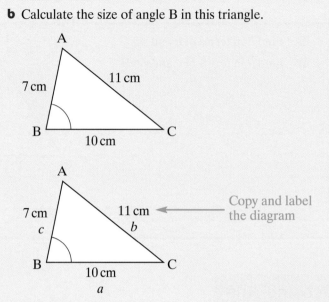

Copy and label the diagram

The cosine rule says

$$a^2 = b^2 + c^2 - 2bc \cos A$$
$$\text{or} \quad b^2 = a^2 + c^2 - 2ac \cos B$$
$$\text{or} \quad c^2 = a^2 + b^2 - 2ab \cos C$$

You need B; you know a, b and c.

$$b^2 = a^2 + c^2 - 2ac \cos B \longleftarrow \quad \text{Select the appropriate formula}$$
$$11^2 = 10^2 + 7^2 - 2 \times 10 \times 7 \cos x$$
$$121 = 100 + 49 - 140 \cos x \longleftarrow \quad \text{Rearranging equation}$$
$$121 = 149 - 140 \cos x \longleftarrow \quad \text{Dividing both sides by 140}$$
$$140 \cos x = 149 - 121$$
$$\cos x = \frac{28}{140}$$
$$\cos x = 0.2$$
$$x = \cos^{-1} 0.2$$
$$x = 78.5° \text{ (1 d.p.)}$$
$$\text{angle B} = 78.5° \text{ (1 d.p.)}$$

The cosine rule works in ANY triangle. You use it when the known facts and the required facts make up three sides and one angle.

Apply 6 🖩

1 Calculate the sides marked x in these triangles.

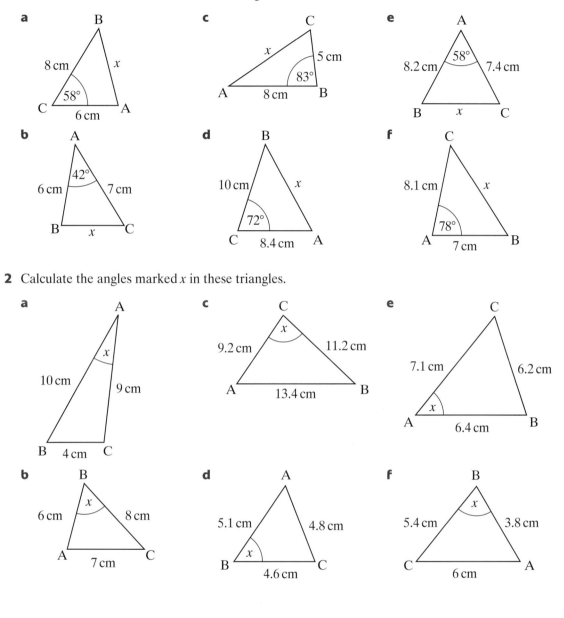

a

B
8 cm
x
58°
C   6 cm   A

b

A
42°
6 cm   7 cm
B x C

c

C
x 5 cm
83°
A   8 cm   B

d

B
10 cm x
72°
C   8.4 cm   A

e

A
58°
8.2 cm   7.4 cm
B x C

f

C
8.1 cm x
78°
A   7 cm   B

2 Calculate the angles marked x in these triangles.

a

A
x
10 cm   9 cm
B   4 cm   C

b

B
x
6 cm   8 cm
A   7 cm   C

c

C
x
9.2 cm   11.2 cm
A   13.4 cm   B

d

A
5.1 cm   4.8 cm
x
B   4.6 cm   C

e

C
7.1 cm   6.2 cm
x
A   6.4 cm   B

f

B
x
5.4 cm   3.8 cm
C   6 cm   A

3 For each part, draw and label a diagram, and use the cosine rule to find the side or angle (marked '?').

	A	B	C	a	b	c
a		71°		11 cm	?	12 cm
b	?			10 cm	14 cm	10 cm
c	?			14 cm	15 cm	19 cm
d			77.5°	11 cm	14 cm	?
e			57°	41 cm	35 cm	?
f			?	11 cm	14.2 cm	12 cm
g		71°		7.1 cm	?	8.4 cm
h		?		12 cm	13 cm	11 cm
i		57°		14.2 cm	?	17 cm
j	?			8.1 cm	9.4 cm	9.4 cm

4 Get Real!

The village of Alton is 7.4 km from Beeville on a bearing of 072°.
Seaford is 9.2 km from Beeville on a bearing of 147°.

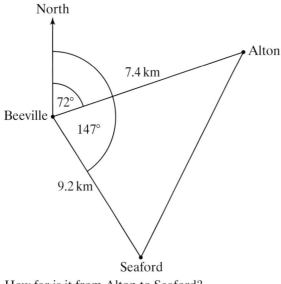

How far is it from Alton to Seaford?

5 Luigi is trying to calculate BC in this triangle.
This is his working.
$a^2 = b^2 + c^2 - 2bc \cos A$
$a^2 = 81 + 49 - 126 \cos 42°$
$a^2 = 130 - 126 \cos 42°$
$a^2 = 4 \cos 42°$
$a^2 = 2.97257930 ...$
$a = 1.7$ cm (to the nearest mm).
Luigi knows this is too small. What has he done wrong?
What is the correct answer?

6 A triangle ABC has AB = 9 cm, BC = 12.2 cm and AC = 7.3 cm.
Calculate angle BAC.

7 Ollie calculates the length of AC in this triangle as follows:
$b^2 = a^2 + c^2 - 2ac \cos B$
$b^2 = 46.24 + 51.84 - 22.02720724 \ldots$
$b^2 = 76.05279275 \ldots$
$b = 8.7$ cm (to 2 s.f.).
Is he correct? If not, find his mistake and correct his answer.

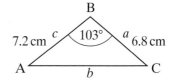

8 Get Real!

Kevin needs to secure a flagpole at an angle of 70° to the horizontal.
The flagpole is 12 m high, and he is using a rope 20 m long.
He fixes one end of the pole with a block, and ties the rope to the other end.
He lifts the pole by pulling the rope.
How far from the foot of the pole does he need to pull the rope
(marked x in the diagram) so that the pole is at the correct angle?
(You may need to calculate other angles first.)

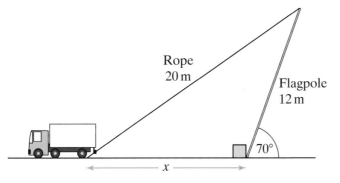

9 Calculate the side and angle marked x in these triangles.

a

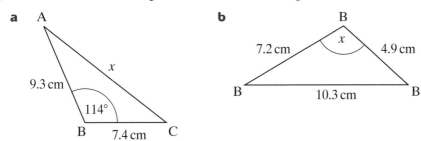

b

Explore

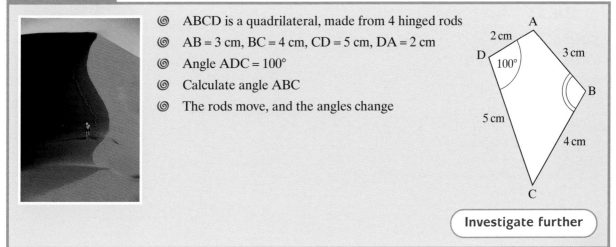

◎ ABCD is a quadrilateral, made from 4 hinged rods
◎ AB = 3 cm, BC = 4 cm, CD = 5 cm, DA = 2 cm
◎ Angle ADC = 100°
◎ Calculate angle ABC
◎ The rods move, and the angles change

Investigate further

Learn 7 Areas of triangles

Example: Find the area of this triangle.

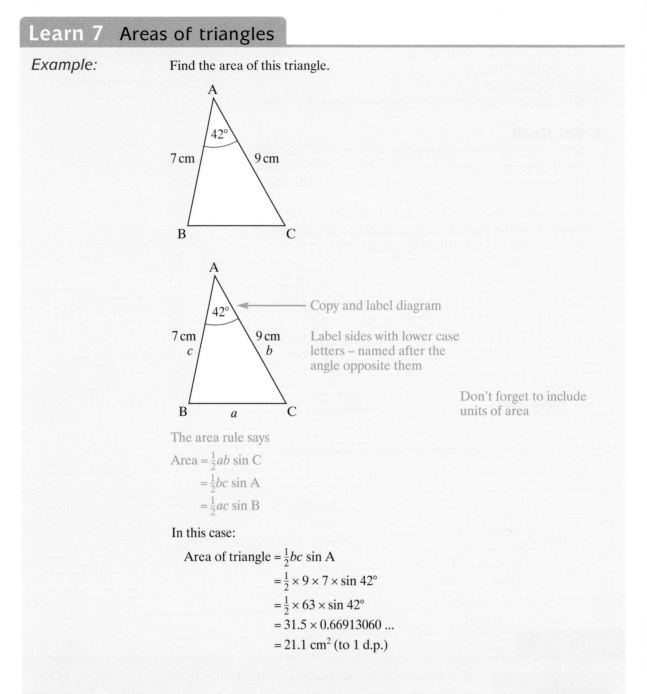

The area rule says

Area $= \frac{1}{2}ab \sin C$

$ = \frac{1}{2}bc \sin A$

$ = \frac{1}{2}ac \sin B$

In this case:

Area of triangle $= \frac{1}{2}bc \sin A$

$ = \frac{1}{2} \times 9 \times 7 \times \sin 42°$

$ = \frac{1}{2} \times 63 \times \sin 42°$

$ = 31.5 \times 0.66913060 \ldots$

$ = 21.1 \text{ cm}^2 \text{ (to 1 d.p.)}$

Apply 7 ⊞

1 Calculate the areas of these triangles.

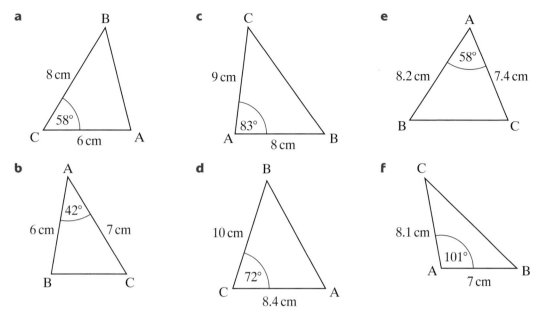

a
B
8 cm
58°
C   6 cm   A

c
C
9 cm
83°
A   8 cm   B

e
A
58°
8.2 cm   7.4 cm
B   C

b
A
42°
6 cm   7 cm
B   C

d
B
10 cm
72°
C   8.4 cm   A

f
C
8.1 cm
101°
A   7 cm   B

2 a Barry says that these two triangles have the same area.

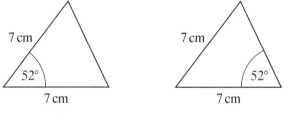

7 cm
52°
7 cm

7 cm
52°
7 cm

Is he right?
Give a reason for your answer.

b Larry says these two triangles have the same area.

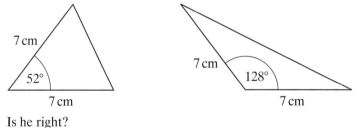

7 cm
52°
7 cm

7 cm
128°
7 cm

Is he right?
Give a reason for your answer.

3 Triangle ABC is isosceles, with AB = BC = 9 cm. Angle BAC = 71°.
Draw a sketch of the triangle, and then calculate its area.

4 Get Real!

A farmer needs to find the area of a field. The field is an irregular shape, so he makes this plan.

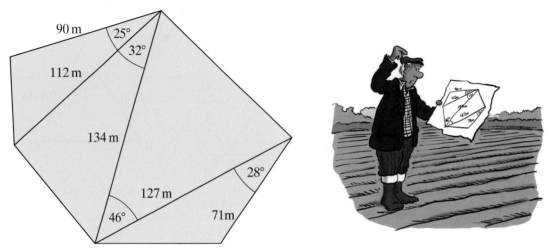

Calculate the area of the field.

5 This triangle has an area of 20 cm².
Calculate the two possible sizes of angle A.

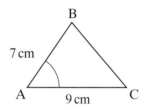

6 An isosceles triangle ABC has AB = AC = 8 cm and angle A = $x°$.
The area of the triangle is 24 cm².
Find all possible values of x.

Explore

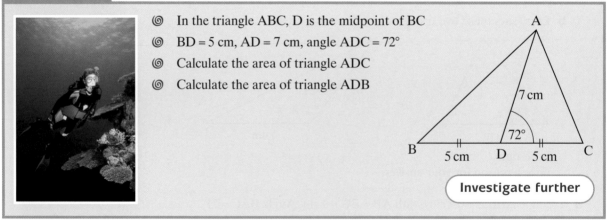

- ◎ In the triangle ABC, D is the midpoint of BC
- ◎ BD = 5 cm, AD = 7 cm, angle ADC = 72°
- ◎ Calculate the area of triangle ADC
- ◎ Calculate the area of triangle ADB

Investigate further

Trigonometry

The following exercise tests your understanding of this chapter, with the questions appearing in order of increasing difficulty.

1 Draw and label a diagram of each triangle ABC below.
Work out the missing sides, angles and areas and copy and complete the table.
Give your answers to 1 d.p.

	Angle A	Angle B	Angle C	Side AB	Side BC	Side CA	Area
a		90°	42°			7 cm	
b	79°	90°		17 cm			
c	55.5°		90°		21.9 cm		
d			90°			32 cm	192 cm²

2 A mast, 1000 m high, is held in place by wires attached to the top of the mast. The wires are at an angle with the ground of 76.1°

  a How long are the wires?

  b How far from the base of the mast are they secured to the ground?

3 A rocket is fired from O at 78.5° to the horizontal. It travels for 7 miles to point A, where it separates into two stages. The second stage flies another 20 miles at 56.3° to the horizontal to point B.

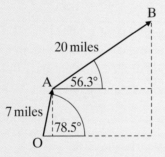

  a Calculate the vertical height of B above the launch pad.

  b Find the total horizontal distance the rocket has travelled down range.

4 Draw and label a diagram for each triangle ABC below.
Work out the missing sides, angles and areas and copy and complete the table.
Give your answers to 1 d.p.

	Angle A	Angle B	Angle C	Side AB	Side BC	Side CA	Area
a	27°		71°	14 cm			
b	72°			19 cm		26 cm	
c	66°					6.9 cm	55.155 cm²

5 The diagram shows a pyramid, ABCD, with vertex O. The base is a rectangle, 16 ft by 12 ft and M is the midpoint of BC. The angle OMN is 62°.

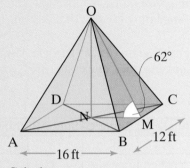

Calculate:

a the height, ON, of the pyramid

b the length of the edge OM

c the length of the edge OA

d the angle between the edge OA and the base of the pyramid.

Give your answers to an appropriate degree of accuracy.

Try some real past exam questions to test your knowledge:

6 a Sketch the graph of $y = \cos x$ for $0° \leqslant x \leqslant 360°$

  b You are given that $\cos 27° = 0.891$

    i Solve the equation $\cos x = 0.891$ for $180° \leqslant x \leqslant 360°$

    ii Solve the equation $\cos x = -0.891$ for $0° \leqslant x \leqslant 360°$

    iii State a solution of the equation $\cos (x - 90°) = 0.891$ for $0° \leqslant x \leqslant 360°$

    iv State a solution of the equation $\sin x = 0.891$ for $0° \leqslant x \leqslant 360°$

Spec A, Higher Paper 1, Nov 04

7 The diagram shows a triangle ABC.
AB = 6 cm, BC = 5 cm and angle B = 75°

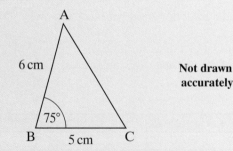

Not drawn accurately

You are given that $\sin 75° = 0.966$ to 3 significant figures.

Calculate the area of the triangle.
Give your answer to an appropriate degree of accuracy.

Spec B, Module 5 Paper 1, June 04

Other functions

B ▶ **Examiners would normally expect students who get a B grade to be able to:**

Complete tables for, and draw graphs of, cubic and reciprocal functions

Use the graphs to solve equations

A* ▶ **Examiners would normally expect students who get an A* grade also to be able to:**

Solve cubic equations by drawing appropriate lines on graphs

Plot and sketch graphs of exponential functions

Recognise the shapes of graphs of functions

What you should already know ...

■ Plot and interpret straight line and quadratic graphs

■ Use quadratic graphs to solve equations

■ Use sine, cosine and tangent to calculate angles and sides of triangles

■ Calculate proportional changes

Cubic functions – functions of the form $f(x) = ax^3 + bx^2 + cx + d$ where a, b, c and d are constants and $a \neq 0$; the simplest cubic function is $f(x) = x^3$ (in this case, a is 1 and the other constants are zero)

The diagram shows the graph of $y = x^3$ and the graph of a more general cubic function, $y = x^3 - x^2 - 2x$

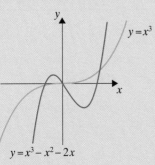

Reciprocal functions – functions which involve negative powers of the variable; the simplest reciprocal function is $y = x^{-1}$, more commonly written as $y = \frac{1}{x}$ or $xy = 1$

This function does not exist for $x = 0$ so its graph has two separate parts, which approach the axes very closely but never meet them

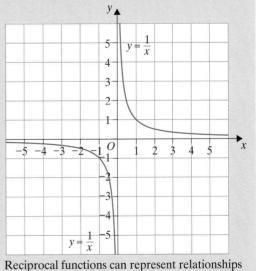

Reciprocal functions can represent relationships of inverse proportion

Exponential functions – functions of the form $y = Ak^{mx}$ where A, k and m are constants, for example, if £1500 is invested at 2.5% compound interest, the amount of money, £y, in the account after x years is given by the formula $y = 1500 \times 1.025^x$, which is an exponential function with $A = 1500$, $k = 1.025$ and $m = 1$

The amount of money in the account is said to **grow exponentially**, which does not necessarily mean that it grows very rapidly, but that its rate of growth is proportional to the amount in the account

Exponential decay – similar to exponential growth, but the amount decreases instead of increasing; the rate of decay is proportional to the amount remaining

The diagram shows graphs of exponential growth and decay; note the similarities and differences

Exponential growth and decay

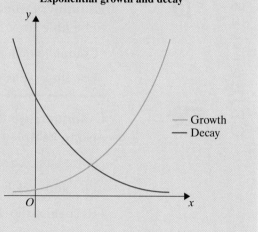

— Growth
— Decay

Learn 1 Cubic functions

Examples:

a Draw the graph of the cubic function $y = x^3 - 2x^2 - 6x$ for $-3 \leqslant x \leqslant 5$

b Use the graph to solve the equations:
  i $x^3 - 2x^2 - 6x = 0$
  ii $x^3 - 2x^2 - 6x = -10$
  iii $x^3 - 2x^2 - 4x = 3$

a Make a table of values.

A calculator is not needed for finding the values in the table

x	−3	−2	−1	0	1	2	3	4	5
x^3	−27	−8	−1	0	1	8	27	64	125
$-2x^2$	−18	−8	−2	0	−2	−8	−18	−32	−50
$-6x$	18	12	6	0	−6	−12	−18	−24	−30
y	−27	−4	3	0	−7	−12	−9	8	45

Draw an *x*-axis labelled from −3 to 5 and a *y*-axis labelled from −30 to 50.

These are the smallest and largest *x* and *y* values in the table

Plot all the points and join them with a smooth curve.

The graph should look like this; it is a typical cubic graph – note the two turning points

On some cubic graphs, the two turning points occur in the same place so that the graph just 'hesitates' and then goes on the in the same direction (see question 2 in Apply 1)

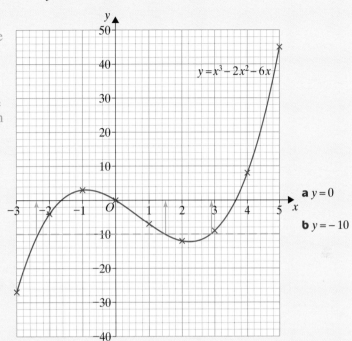

$y = x^3 - 2x^2 - 6x$

a $y = 0$

b $y = -10$

b i The equation $x^3 - 2x^2 - 6x = 0$ is solved by finding the values of *x* where the graph cuts the *x*-axis (that is, finding the values of *x* for which $y = 0$). The graph cuts the *x*-axis at three points, so the equation $x^3 - 2x^2 - 6x = 0$ has three solutions, $x = -1.6$, $x = 0$ and $x = 3.6$ correct to 1 d.p.

If *x* has one of these values, $x^3 - 2x^2 - 6x$ is equal to zero

ii To solve the equation $x^3 - 2x^2 - 6x = -10$, find the values of *x* for which $y = -10$ by drawing the line $y = -10$ and finding where it cuts the graph. Again, there are three solutions, $x = -2.4$, $x = 1.5$ and $x = 2.9$ correct to 1 d.p.

If *x* has one of these values, $x^3 - 2x^2 - 6x$ is equal to −10

iii The equation $x^3 - 2x^2 - 4x = 3$ is also solved by drawing a line on the diagram and finding where it cuts the graph.

Compare the equation of the graph with the equation to be solved.

Equation of graph:

$x^3 - 2x^2 - 6x = y$

The left-hand side of the equation to be solved has to be made the same as the left-hand side of the equation of the graph

Equation to be solved:

$x^3 - 2x^2 - 4x = 3$

Note that the only difference is that the equation of the graph has −6*x* and the equation to be solved has −4*x*

Subtract 2*x* from both sides of the equation to be solved.

$x^3 - 2x^2 - 6x = 3 - 2x$

The solution of the equation is the value (or values) of *x* for which $x^3 - 2x^2 - 6x$ and $3 - 2x$ are the same

Draw the graph of $y = 3 - 2x$ and see where it meets the cubic graph $y = x^3 - 2x^2 - 6x$.

This value (or these values) are the *x*-coordinate(s) of the point(s) of intersection of the graphs of $y = x^3 - 2x^2 - 6x$ and $y = 3 - 2x$

261

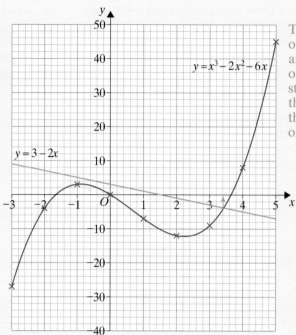

$y = x^3 - 2x^2 - 6x$

$y = 3 - 2x$

This equation has only one solution as the line and the curve cross at just one point, but a different straight line could meet the cubic graph at two or three places, giving two or three solutions

The two graphs meet at the point where x is approximately 3.4, so the solution of the equation $x^3 - 2x^2 - 4x = 3$ is $x = 3.4$ correct to 1 d.p.

Apply 1

1 For this question, draw the graph of $y = x^3 - 2x^2 - 6x$ in the example above.

   a Solve the equations:

      i $x^3 - 2x^2 - 6x = 2$ **iii** $x^3 - 2x^2 - 6x = 10$

      ii $x^3 - 2x^2 - 6x = 5$ **iv** $x^3 - 2x^2 - 6x = -20$

   b Solve the equations:

      i $x^3 - 2x^2 - 6x = x$ **ii** $x^3 - 2x^2 - x = 0$

   c Explain why some of the equations have three solutions and some have only one.

   d Find an equation that has two solutions.

2 The diagram shows the graph of $y = -x^3 + 2x^2 + 11x - 9$

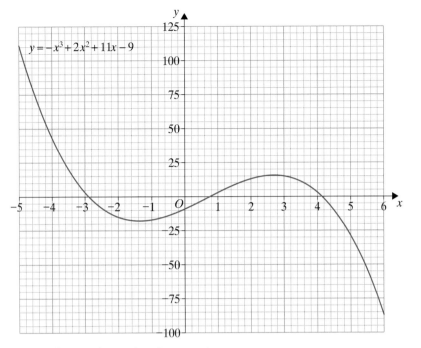

$y = -x^3 + 2x^2 + 11x - 9$

a Use the graph to solve the equations:

　i $-x^3 + 2x^2 + 11x - 9 = 0$

　ii $-x^3 + 2x^2 + 11x - 9 = 25$

　iii $-x^3 + 2x^2 + 11x - 9 = -10$

　iv $-x^3 + 2x^2 + 11x - 9 = x$

b Find another equation that has three solutions.

c Find another equation that has only one solution.

d Find an equation that has two solutions.

3 On one diagram, draw the graphs of $y = x^3$ and $y = -x^3$ for values of x from −3 to 3.

a What is the same and what is different about the two graphs?

b Which point is on both of the graphs?

4 On one diagram, draw the graphs of $y = x^3$ and $y = \frac{1}{2}x^3$ for values of x from −3 to 3.

a What is the same and what is different about the two graphs?

b Draw the line $y = 12$ on the diagram and use it to solve the equations:

　i $x^3 = 12$ 　　　　　　　**ii** $\frac{1}{2}x^3 = 12$

5 Karl makes a table of values for $y = x^3 + 2x$ and draws the graph below.

x	−2	−1	0	1	2	3
x^3	−8	−1	0	1	8	27
$2x$	−4	−2	0	2	4	6
y	−12	−3	0	3	12	33

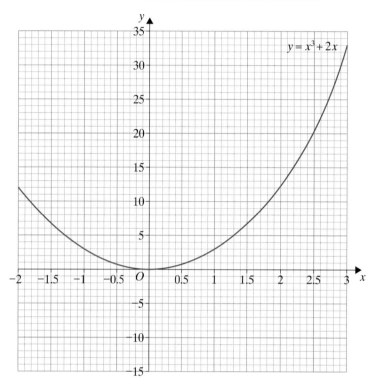

Has Karl completed the table and drawn the graph correctly?
Give a reason for your answer.

6 This table of values is for the function $f(x) = x^3 - 5x - 2$ for $-4 \leqslant x \leqslant 4$

x	−4	−3	−2	−1	0	1	2	3	4
x^3	−64								
$-5x$	20								
-2	−2	−2	−2	−2	−2	−2	−2	−2	−2
y	−46								

a Copy and complete the table of values.

b Draw suitable axes on graph paper and draw the graph of
$f(x) = x^3 - 5x - 2$

c i Write the x-coordinates of the points where the graph cuts the
x-axis.

ii For which equation are these x-values the solution?

d Use the graph to solve the equation $x^3 - 5x - 2 = -2$

e Solve the equation $x^3 - 2x = 0$ by drawing a line on the graph

7 Get Real!

A rectangular box is constructed from a square of card measuring 30 cm by 30 cm. A square measuring x cm by x cm is cut from each corner and the resulting flaps are folded up to make the sides of the box.

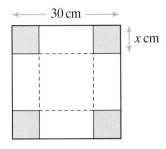

a Explain why the volume of the box, V cm^3, is given by the cubic function $V = x(30 - 2x)^2$

b i Use the equation $V = x(30 - 2x)^2$ to show that the values of x for which $V = 0$ are 0 and 15.

   ii Use the diagram to explain why the volume is zero when the length of the sides of the cut out squares is 0 cm or 15 cm.

c i Draw the graph of the equation $V = x(30 - 2x)^2$ for values of x from 0 to 15.

   ii Use the graph to find the maximum possible volume of the box and the value of x that gives the maximum volume.

8 Get Real!

Part of a roller coast track fits the graph of the function

$y = -\dfrac{x^3}{3000} + \dfrac{3x^2}{40} - 5x + 120$, where the horizontal distance is x feet and

the corresponding height is y feet. The diagram shows the part of the graph of this function.

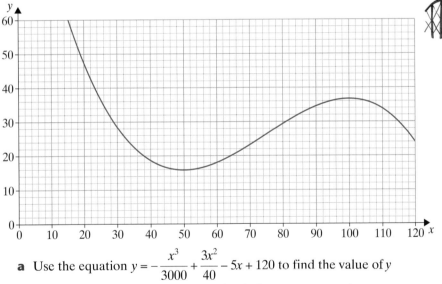

a Use the equation $y = -\dfrac{x^3}{3000} + \dfrac{3x^2}{40} - 5x + 120$ to find the value of y

when x is 50 and use the graph to check that your answer is correct.

b Use the graph to find:

   i the height of the ride when the horizontal distance is 20 feet

   ii the horizontal distances when the height of the ride is 30 feet.

c Explain why the cubic function would not be a good model for the whole of the roller coaster ride.

Explore

- The general form of a cubic function is $y = ax^3 + bx^2 + cx + d$, where a, b, c and d are constants
- Starting with b, c and d all equal to zero, choose some simple values of a. Investigate the effect of varying the value of a on the graph of the function; consider both positive and negative values of a
- A graphic calculator or computer graph-drawing program will be useful

Investigate further

Learn 2   Reciprocal functions

Examples:

a Draw the graph of the reciprocal function $y = \frac{1}{x}$ for $-6 \leqslant x \leqslant 6$

b Use the graph to solve the equation $\frac{1}{x} = x + 3$

a Make a table of values, draw suitable axes and draw the graph.

x	−6	−4	−2	−1	−0.1	0	0.1	1	2	4	6
y	−0.17	−0.25	−0.5	−1	−10	−	10	1	0.5	0.25	0.17

It is not possible to divide by zero so there is no value for y when $x = 0$

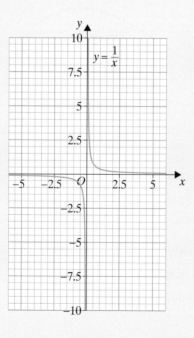

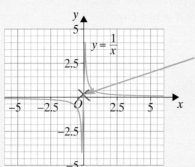

There is a break in the graph between the very large negative y-values when x is small and negative and the very large positive y-values when x is small and positive

Do not make the mistake of trying to join up the two parts of the graph

The graph is discontinuous

b To solve the equation $\frac{1}{x} = x + 3$, draw the line $y = x + 3$ and find the values of x where it cuts the graph of $y = \frac{1}{x}$

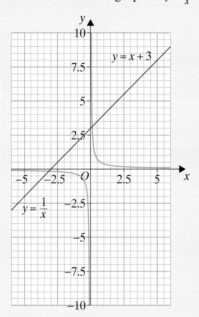

The straight line cuts the reciprocal graph in two places, so the equation $\frac{1}{x} = x + 3$ has two solutions. The solutions are $x = -3.3$ and $x = 0.3$

Apply 2

Apart from question 6 this is a calculator exercise.

1 a Copy and complete this table of values for the reciprocal function $y = \frac{2}{x}$

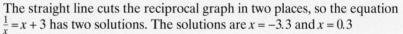

x	−20	−15	−10	−5	−2	−1	−0.1	0	0.1	1	2	5	10	15	20
y	−0.1						−20	−							

b Draw the graph of $y = \frac{2}{x}$

c Use your graph to solve the equations:

   i $\frac{2}{x} = 2$ **iii** $\frac{2}{x} = 4$

   ii $\frac{2}{x} = -2$ **iv** $\frac{2}{x} = -4$

2 The diagram shows the graph of the reciprocal function $y = \frac{12}{x} + 3$

You will need a calculator to do the last part of this question.

  a Use the graph to solve the equations

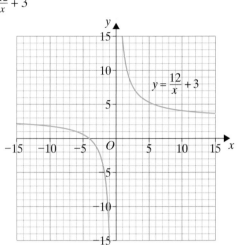

$$y = \frac{12}{x} + 3$$

  i $\frac{12}{x} + 3 = 5$ **iv** $\frac{12}{x} + 3 = \frac{1}{2}x$

  ii $\frac{12}{x} + 3 = x$ **v** $\frac{12}{x} + 3 = 3x$

  iii $\frac{12}{x} + 3 = 2x$

  b i Explain why the equation $\frac{12}{x} + 3 = -x$ does not have any solutions.

  ii Find another equation like this that does not have any solutions.

  c Solve the equations in part **a** using algebra instead of the graph.

3 Here is the table of values for the equation $y = \frac{10}{x} - 2$

x	−10	−8	−6	−4	−2	0	2	4	6	8	10
$\frac{10}{x}$	−1						5				1
−2	−2	−2	−2	−2	−2	−2	−2	−2	−2	−2	−2
y							3				

  a Copy the table and complete it.

  b Use the table to plot the graph and join the two separate parts with smooth curves.

  c Use the graph of $y = \frac{10}{x} - 2$ to solve the equations:

  i $\frac{10}{x} - 2 = 4$

  ii $\frac{10}{x} - 2 = -4$

  iii $\frac{10}{x} - 2 = x$

  d The equation $\frac{10}{x} - 2 = 4$ in part **c i** simplifies to $\frac{10}{x} = 6$

  i Solve this equation algebraically.

  ii Find simpler versions of the equations in parts **c ii** and **c iii** and solve them algebraically.

  iii Solve the equation $\frac{10}{x} - 5 - x = 0$ both algebraically and using the graph of $y = \frac{10}{x} - 2$. Check that your answers agree.

  e How would the graph of the equation $y = \frac{10}{x}$ compare with the graph of the equation $y = \frac{10}{x} - 2$?

  f How would the graph of the equation $y = \frac{10}{x} + 2$ compare with the graph of the equation $y = \frac{10}{x} - 2$?

4 a On the same diagram, draw the three graphs with equations $y = \frac{1}{x}$, $y = \frac{1}{x} + 1$ and $y = \frac{1}{x-1}$ for $-10 \leqslant x \leqslant 10$

b Write what is the same and what is different about the graphs.

5 On the same diagram, draw the graphs of $y = \frac{1}{x}$ and $y = \frac{1}{x^2}$ for values of x from -10 to 10.

a What are the similarities and what are the differences between the two graphs?

b Explain why there are no negative y-values for the graph of $y = \frac{1}{x^2}$

6 A rectangle has an area of 36 cm^2.
Its width is x cm and its height is y cm.

a Find an equation connecting x and y.

b Explain why x and y cannot be negative in this case.

c Draw a graph of y against x for $0 < x \leqslant 36$

d Use your graph to find the width of the rectangle when its height is 7 cm.

e Which of these statements are correct?

  i The product of x and y is constant

  ii The sum of x and y is constant

  iii x is inversely proportional to y

  iv y is inversely proportional to x

7 Get Real!

The time taken to complete a journey of a fixed distance is inversely proportional to the average speed of the journey. For a car journey of 250 miles this relationship is represented by the reciprocal function $v = \frac{250}{t}$, where v mph is the average speed for the journey and t hours is the time taken.

a What ranges of values for v and t would be appropriate for this situation?

b Sketch a graph of $v = \frac{250}{t}$ for your chosen values of v and t.

8 Get Real!

The specification for each cylinder of a 4-cylinder car engine requires that the capacity is 0.4 litres (400 cc), to give a total engine capacity of 1.6 litres. The size of the cylinder depends upon its bore (the diameter) and its stroke (the height).

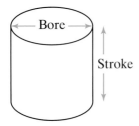

a The formula for the volume of a cylinder is $V = \pi r^2 h$. Show that, for the 400 cc car cylinder, $s = \frac{1600}{\pi b^2}$, where s cm is the stroke and b cm is the bore.

b Draw a graph of this reciprocal function for values of b from 5 to 20.

c Use your graph to find:

  i the stroke when the bore is 17 cm

  ii the bore when the stroke is 10 cm.

d A car cylinder is said to be 'square' when the bore is equal to the stroke.

  i Use your graph to find the bore and the stroke when the cylinder is 'square'.

  ii Use the equation $s = \frac{1600}{\pi b^2}$ to find the answer algebraically.

Learn 3 Exponential functions

Examples:

a A simple example of an exponential function is $y = 2^x$
  i Find the values of y when x is 2, 3, 0, −1, −5.
  ii Complete a table of values for the function $y = 2^x$ for $-5 \leqslant x \leqslant 5$
  iii Draw a graph of the function $y = 2^x$ for $-5 \leqslant x \leqslant 5$
  iv Use the graph to solve the equation $2^x = 10$

a **i** When $x = 2, y = 2^2 = 2 \times 2 = 4$
When $x = 3, y = 2^3 = 2 \times 2 \times 2 = 8$
Remember that any number to the power $0 = 1$
When $x = 0, y = 2^0 = 1$
When $x = -1, y = 2^{-1} = \frac{1}{2} = 0.5$ and,
when $x = -5, y = 2^{-5} = \frac{1}{2^5} = \frac{1}{32} = 0.03125$

Any of these answers can be found by using the y^x or x^y or $\hat{\ }$ calculator button; make sure you know which button it is on your calculator and how to use it

ii Use the results from part **a** in the table of values and find the other answers in a similar way.

x	−5	−4	−3	−2	−1	0	1	2	3	4	5
y	0.03	0.06	0.13	0.25	0.5	1	2	4	8	16	32

iii

It looks as if the graph of $y = 2^x$ meets the x-axis when $x = -3$

In fact, the graph gets closer and closer to the x-axis but never actually touches it

y gets closer and closer to zero but never actually becomes zero (see the magnified version of part of the graph below)

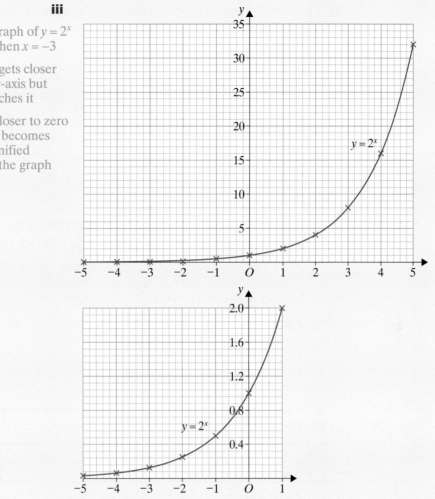

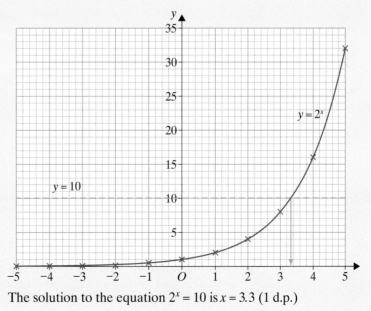

iv To solve the equation $2^x = 10$, draw the line $y = 10$ and find the value of x where it cuts the graph of $y = 2^x$

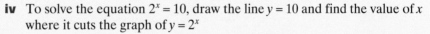

The solution to the equation $2^x = 10$ is $x = 3.3$ (1 d.p.)

b A laboratory culture of bacteria increases by 10% each hour. Initially there are 500 bacteria in the culture.

   i Show that the number, y, of bacteria after x hours, is $y = 500 \times 1.1^x$

   ii Use the equation to calculate the number of bacteria after 5 hours.

   iii Draw a graph of y against x for $0 \leqslant x \leqslant 10$

   iv Use the graph to find the number of bacteria after $7\frac{1}{2}$ hours and the time taken for the number of bacteria to double.

> More complex exponential functions can be used to describe proportional growth and decay such as money gaining interest in the bank, depreciation in value of cars as they get older, decay of radioactive substances and growth in populations of bacteria

b i After one hour, the number of bacteria is 110% of the number at the start. So after one hour there are $500 \times \frac{110}{100} = 500 \times 1.1$ bacteria.

After a further hour, the number of bacteria has again increased by 10%, so the number has been multiplied again by 1.1. So, after two hours, the number of bacteria is $500 \times 1.1 \times 1.1 = 500 \times 1.1^2$

For every hour that passes, the number of bacteria is multiplied by 1.1

Hence, after x hours, the number of bacteria is 500×1.1^x

> The general form of the exponential function is $f(x) = Ak^{mx}$
>
> For the bacteria, A is 500, k is 1.1 and m is 1

ii The number of bacteria after 5 hours is $500 \times 1.1^5 \approx 805$

iii The table of values for the graph is:

x	0	1	2	3	4	5	6	7	8	9	10
y	500	550	605	666	732	805	886	974	1072	1179	1297

The graph is:

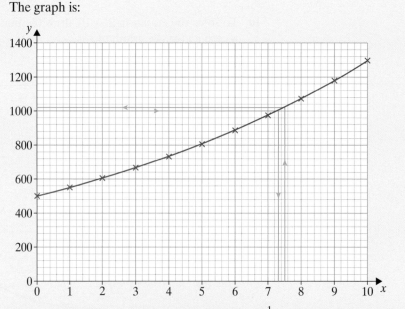

iv From the graph, the number of bacteria after $7\frac{1}{2}$ hours is approximately 1020. The time taken for the number to double is approximately 7.3 hours, i.e. approximately 7 hours 20 minutes.

Apply 3

1 Work out:

a $2^4, 2^{-2}, 2^0, -2^2, (-2)^2$

b $1^4, 1^{-2}, 1^0, -1^2, (-1)^2$

c $10^4, 10^{-2}, 10^0, -10^2, (-10)^2$

d $0.1^4, 0.1^{-2}, 0.1^0, -0.1^2, (-0.1)^2$

e $(\frac{1}{2})^4, (\frac{1}{2})^{-2}, (\frac{1}{2})^0, -(\frac{1}{2})^2, (-\frac{1}{2})^2$

2 a Draw a graph of the exponential function $y = 4^x$ for values of x from -3 to 3.

b Use the graph to:

  i estimate the value of $4^{2.5}$ and $4^{-1.5}$

  ii solve the equations $4^x = 5$ and $4^x = 2$

3 A culture of bacteria increases by 10% each hour. Initially there are 500 bacteria.

a Use the formula $y = 500 \times 1.1^x$ to find:

  i the number of bacteria after three and a half hours

  ii the number of bacteria after twenty minutes

  iii the increase in the number of bacteria in the fourth hour.

b Use the graph of $y = 500 \times 1.1^x$ to estimate:

  i the time at which the number of bacteria reaches 1000

  ii the time taken for the number of bacteria to increase from 1000 to 1200.

c What is the percentage increase in the number of bacteria from the beginning of the experiment to 24 hours later?

4 Another culture of bacteria increases by 20% each hour and there are initially 1000 bacteria.

a Find a formula for the number of bacteria, y, after x hours.

b Draw a graph of y against x for $0 \leqslant x \leqslant 10$.

c Use your graph to find:

  i the amount of time taken for the number of bacteria to double from 1000 to 2000

  ii the amount of time taken for the number of bacteria to double from 2000 to 4000.

d Jean says that if the number of bacteria increases by 20% an hour, the percentage increase after 5 hours is 100%. Anne says that the percentage increase will be 149%.
Who is correct? Give a reason for your answer.

5 The general form of exponential functions is $f(x) = Ak^{mx}$
Identify the values of A, k and m in each of these exponential functions.

a $f(x) = 25 \times 1.5^{2x}$

b $f(x) = 2^{-x}$

c $f(x) = \dfrac{5}{1.02^x}$

 6 This is a sketch of the graph of $y = 2^x$ for values of x from -1 to 5.

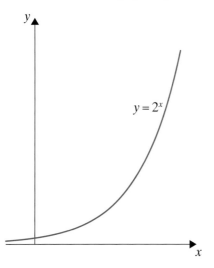

a What are the coordinates of the point where the graph cuts the y-axis?

b Copy the diagram and sketch on it the graphs of these equations:

i $y = 2^{2x}$

ii $y = -2^x$

iii $y = 2^{-x}$

iv $y = 2^x + 5$

7 Get Real!

At the beginning of 2006, £1000 is invested in a bank account paying 4% a year interest; the interest is added on at the end of each year.

a Show that the amount in the account after x years, £y, is given by the equation $y = 1000 \times 1.04^x$

b i Use the equation to calculate the amount in the account after 5 years.

ii By what percentage has the amount in the account increased after 5 years?

c Draw a graph of $y = 1000 \times 1.04^x$ for $0 \leqslant x \leqslant 10$.

d Explain why the graph does not correctly represent the amount of money in the account after, for example, four and a half years.

8 Get Real!

A radioactive isotope of arsenic, Arsenic 75, loses mass rapidly by radioactive decay. If the mass of Arsenic 75 initially is 10 kg, the approximate mass, y kg, remaining after x days is given by the equation $y = 10 \times 0.96^x$

a Use this formula to find the mass of Arsenic 75 remaining after 10 days.

b Use the trial and improvement method to find how many days it takes for the mass remaining to be less than 1 kg.

c Draw the graph of $y = 10 \times 0.96^x$ for $0 \leqslant x \leqslant 20$.

d i Use your graph to find how long it takes for the mass of Arsenic 75 to be halved from 10 kg to 5 kg.

   ii Show that after the same time again the mass has halved once more.

e What feature of the equation $y = 10 \times 0.96^x$ indicates that y decreases as x increases?

f The equation $y = 10 \times 0.96^x$ is equivalent to the equation $y = 10 \times k^{-x}$ Find the value of k.

g i How would the graph of $y = 20 \times 0.96^x$ compare with the graph of $y = 10 \times 0.96^x$?

   ii How would the graph of $y = 10 \times 0.56^x$ compare with the graph of $y = 10 \times 0.96^x$?

Other functions

ASSESS

The following exercise tests your understanding of this chapter, with the questions appearing in order of increasing difficulty.

1 a Copy and complete the table of values for $y = x^3 - 2x^2 - 5x + 6$

x	−3	−2	−1	0	1	2	3	4
x^3	−27	−8						
$-2x^2$	−18	−8						
$-5x$	+15	+10						
$+6$	+6	+6	+6					
y	−24							

b Draw the graph of $y = x^3 - 2x^2 - 5x + 6$

c Use your graph to solve the equation $x^3 - 2x^2 - 5x + 6 = 0$

2 a On the same axes and scales draw the graphs of $y = \sin x$ and $y = \sin 2x$ for values of x from 0° to 360°.

b Find the values of x where $\sin 2x = \sin x$

c Use your graphs to solve these equations:

   i $\sin x = \frac{1}{2}$

   ii $\sin 2x = \frac{1}{2}$

   iii $\sin x = -0.1$

   iv $\sin 2x = -0.9$

3 a Copy and complete the table of values for $y = x^2 + \frac{8}{x}$

x	−4	−3	−2	−1	−0.5	0.5	1	2	3	4
x^2	16	9			0.25					
$+\frac{8}{x}$	−2	−2.7			−16					
y	14									

b Draw the graph of $y = x^2 + \frac{8}{x}$

c Use your graph to solve these equations:

i $x^2 + \frac{8}{x} = 0$

ii $x^2 + \frac{8}{x} = 2$

iii $x^2 + \frac{8}{x} + 3 = 0$

d i On the same diagram, draw the graph of $y = x$.

ii Find the value of x at the point where the line crosses $y = x^2 + \frac{8}{x}$

iii Write down the equation to which this value is the solution.

4 When a new motorway was opened the average rate of traffic flow was measured as 3000 vehicles per hour. Planners predicted that the average rate would rise by 8% per year.

a Write down, in the form, $3000 \times a^n$, where a is a constant, and n is the number of years, the average rate of traffic flow after 1, 2, 3 and 5 years and hence show that the average rate of traffic flow after n years is 3000×1.08^n

b Draw the graph of $y = 3000 \times 1.08^n$ for the first 15 years and from it find the number of years before the average rate of traffic flow doubles.

Try a real past exam question to test your knowledge:

5 The graph shows four curves A, B, C and D.

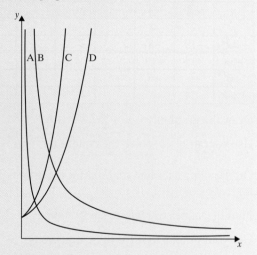

Match each curve to its equation.

a $y = \frac{1}{x}$ is curve ....

c $y = 3^x$ is curve ....

b $y = 2^x$ is curve ....

d $y = \frac{4}{x}$ is curve ....

Spec A, Higher Paper 1, June 05

18 Loci

What you should already know …

- Measure a line accurately (within 2 millimetres)
- Measure and draw an angle accurately (within 2 degrees)
- Construct the perpendicular bisector of a line
- Construct the bisector of an angle
- Construct and interpret a scale drawing
- The equation of a straight-line graph, $y = mx + c$

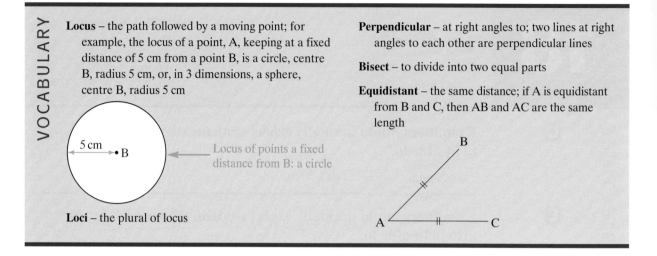

Locus – the path followed by a moving point; for example, the locus of a point, A, keeping at a fixed distance of 5 cm from a point B, is a circle, centre B, radius 5 cm, or, in 3 dimensions, a sphere, centre B, radius 5 cm

Locus of points a fixed distance from B: a circle

Loci – the plural of locus

Perpendicular – at right angles to; two lines at right angles to each other are perpendicular lines

Bisect – to divide into two equal parts

Equidistant – the same distance; if A is equidistant from B and C, then AB and AC are the same length

Learn 1 Constructing loci

Examples:

Draw accurately:

a the locus of points 1 cm from a point A

b the locus of points 1 cm from a line AB

c the locus of points the same distance from two points A and B

d the locus of points the same distance from two lines, AB and AC.

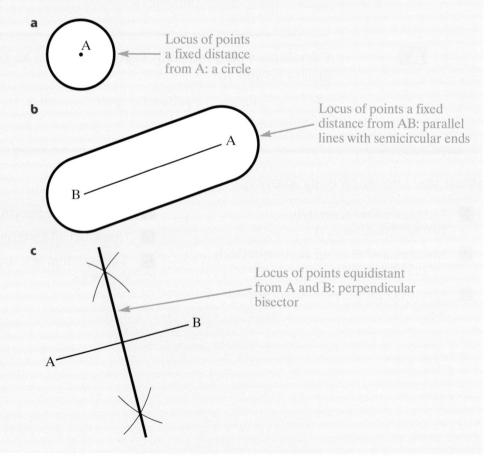

a Locus of points a fixed distance from A: a circle

b Locus of points a fixed distance from AB: parallel lines with semicircular ends

c Locus of points equidistant from A and B: perpendicular bisector

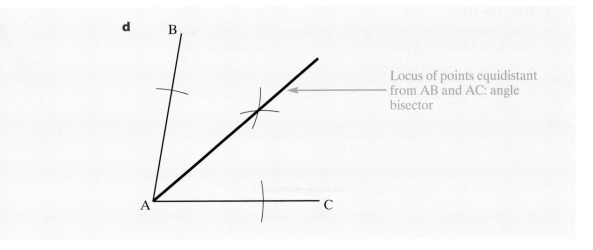

Locus of points equidistant from AB and AC: angle bisector

Apply 1

1 Explore these loci practically.

   a Find the path followed by the corner of a square as the square is rotated about an opposite corner.
     Draw the locus.

   b Find the path of a corner of a square of card as it is rotated against a ruler.
     Draw the locus.

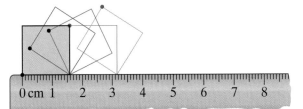

   c Find the locus of a mark on the edge of a coin as it rolls along a ruler.

2 A square has sides of length 6 cm.
Sam says that if an ant moves round the outside of the square,
always staying 3 cm from the square, it walks in a square with sides
of length 12 cm.
Is Sam correct?
Give a reason for your answer.

3 Get Real!

You are going to find the path of a man, halfway up a ladder, as the foot of the ladder slips until the ladder is resting on the ground.

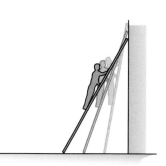

a Cut a strip of card, 10 cm long, to represent the ladder.

b Make a small hole exactly in the middle to represent where the man is.

c Draw two lines at right angles, each at least 10 cm long, to represent the ground and the wall.

d Put the ladder almost vertical, leaning against the wall, and mark the position of the man through the hole in the card.

e Move the ladder as if it has slipped slightly, and again mark where the man is.

f Repeat until the ladder is horizontal.

g Join up the marks you have made to complete the locus.

h Draw the locus.

i What would happen if the man was three quarters of the way up the ladder? Make a new hole in your card 'ladder' and repeat the experiment. Draw the locus.

4 Get Real!

A goat lives in a field 12 m long and 8 m wide. In the centre of the field is a post. The goat is tethered to the post with a rope that is 4 m long.

Using a scale of 1 cm to represent 1 m:

a make a scale drawing of the field

b mark the position of the post

c shade in the area of the field that the goat can reach.

5 a Construct a square ABCD with all sides 8 cm long.

b Construct the locus of all points the same distance from AB and AD.

c Construct the locus of all points 3 cm from A.

d Shade in all the points that are closer to AB than AD and less than 3 cm from A.

6 Get Real!

Tommy builds a toy train track. He draws a rectangle measuring 8 m by 6 m.

He builds the track outside the rectangle, so that the track is always exactly 1 m from the rectangle.

Using a scale of 1 cm to represent 1 m, make a drawing of the rectangle and the track.

7 Construct an equilateral triangle with sides of 7 cm.

Draw the locus of a point that is always 1 cm from the sides of the triangle.

8 Get Real!

A large island has two radio transmitters, M and N, marked × on the diagram.

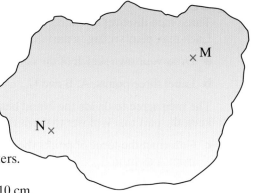

a Draw your own island, and mark two transmitters as in the diagram, so that M and N are 6 cm apart.

b Each transmitter has a range of 5 km. Using a scale of 1 cm to represent 1 km, mark all the points that are in range of transmitter M.

c Shade in the area that is in the range of both transmitters.

9 a Construct an isosceles triangle ABC, with AB = AC = 10 cm and BC = 7 cm

b Shade in all the points inside the triangle that are closer to AB than BC, and closer to B than A.

10 Mark two points A and B, which are 8 cm apart.
Sketch the locus of all points that are closer to A than B, but less than 6 cm from B.

11 Get Real!

A rectangular lawn is 12 m long and 10 m wide.
A gardener has three sprinklers, which spray water in a circle with a radius of 3 m.
She puts a sprinkler at A, which is 3 m from the top edge and 3 m from the left edge; she puts another at B, which is 3 m from the top edge and 3 m from the right edge; and the third at C, which is 3 m from the bottom edge and 6 m from the left edge.
Make a scale drawing of the lawn, and shade in the part that gets watered by the sprinklers.

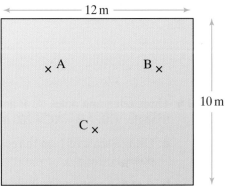

12 Get Real!

A field measures 24 m long and 16 m wide. In one corner, there is a shed, 8 m long and 6 m wide.

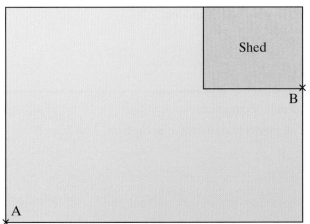

a Using a scale of 1 cm to 2 m, make a scale drawing of the field and the shed.

b A goat is tethered to corner A with a rope, 14 m long.
Using the same scale, shade the area that the goat can reach.

c After a week, the goat is tethered to point B, where the shed meets the edge of the field.
Shade the area the goat can reach now.

<u>13</u> **Get Real!**

Three men discover an island. Each wants to claim it as his own.
Each man plants a flag in part of the island.

a Draw your own sketch of the island.

b Label three points, A, B and C.

The men agree to divide the island between them, so that they
keep the part that is closer to them than to the others.

c Construct the locus of points the same distance
from A as from B.

d Construct the locus of points equidistant
from A and C.

e Construct the locus of points equidistant
from B and C.

f Shade in all points closer to A than B or C.

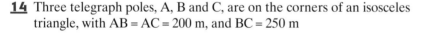

<u>14</u> Three telegraph poles, A, B and C, are on the corners of an isosceles
triangle, with AB = AC = 200 m, and BC = 250 m

a Using a scale of 1 cm to represent 20 m, make a scale drawing of the
poles.

b A man is planning to build a house somewhere within the triangle.
He needs to be at least 90 m away from all telegraph poles.
Shade the area on the plan where he can build his house.

c He wants the house to be rectangular, 50 m wide and 70 m long.
Can he fit a house this size in the shaded area?

Explore

◎ A and B are two points, 9 cm apart

◎ Find the locus of all points that are twice as far from A as from B

Investigate further

Explore

◎ A is a point 8 cm from a line BC

◎ Find the locus of all points that are twice as far from BC as they are from A

Investigate further

Learn 2   Loci and graphs

Examples:

a Find the equation of this circle.

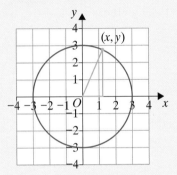

$x^2 + y^2 = r^2$ is the equation of a circle, centre the origin and radius r

The circle has a radius of 3 units.
So the equation is $x^2 + y^2 = 9$

b **i** Draw the graphs of $x^2 + y^2 = 25$ and $y = 3x - 2$
   ii Use your graph to solve the simultaneous equations $x^2 + y^2 = 25$ and $y = 3x - 2$

   i $x^2 + y^2 = 25$ is a circle, centre the origin; $r^2 = 25$, so $r = 5$
         $y = 3x - 2$ has an intercept on the y-axis of -2, with a gradient of 3

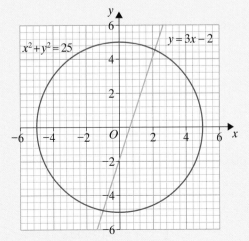

   ii   From the graph, the solutions are $(-1.1, -4.9)$ and $(2.3, 4.4)$

283

Apply 2

1 Construct graphs of these loci.

a $x^2 + y^2 = 16$ **e** $y^2 = 36 - x^2$

b $x^2 + y^2 = 4$ **f** $x^2 = 1 - y^2$

c $x^2 + y^2 = 64$ **g** $81 - x^2 = y^2$

d $x^2 + y^2 = 49$ **h** $x^2 + y^2 = 18$

2 Write the equation of the circle if A is the point:

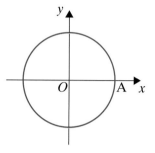

a $(6, 0)$

b $(11, 0)$

c $(20, 0)$

d $(2.5, 0)$

3 Bill says that the graph of $x^2 + y^2 = 4$ passes through the point $(2, 2)$.
Is he correct? Give a reason for your answer.

4 Jeremy says that the graph of $x^2 + y^2 = 10$ passes through the point $(3, -1)$.
Is he correct? Give a reason for your answer.

5 a Draw a circle, centre $(0, 0)$, which passes through $(-3, 4)$.

b What is the radius of the circle?

c What is the equation of the circle?

6 Ed says that the equation of the circle on the right is $x^2 + y^2 = 6$
Is he correct? Give a reason for your answer.

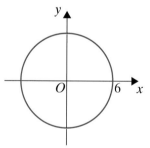

7 What is the equation of the circle, centre $(0, 0)$, which passes through $(-2, 5)$?

8 Write the equation of this circle if A is the point:

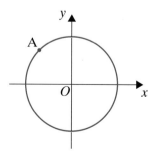

a $(-4, 2)$

b $(-3, 3)$

c $(-\sqrt{2}, \sqrt{3})$

9 By drawing graphs, solve these pairs of simultaneous equations.

a $x^2 + y^2 = 9$ and $y = 2x$ **d** $x^2 + y^2 = 36$ and $y = 4 - 2x$

b $x^2 + y^2 = 25$ and $y = 2x + 1$ **e** $x^2 + y^2 = 16$ and $y = 2x + 3$

c $x^2 + y^2 = 4$ and $y = x + 1$ **f** $x^2 + y^2 = 10$ and $x - y = 4$

10 Get Real!

On the right is a diagram of a football pitch.
The centre circle has a radius of 10 yards.

a Taking the centre spot as the origin, write the equation of the centre circle.

A defender at $(0, -12)$ passes the ball to a forward at $(8, 4)$.

b By drawing a graph, find:

i the point where the ball crossed the centre circle

ii the equation of the line that represents the pass from $(0, -12)$ to $(8, 4)$.

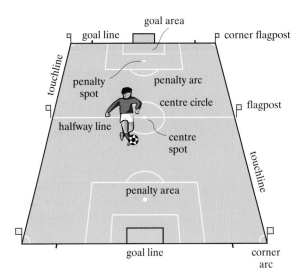

11 A circle, centre the origin, is crossed by a straight line at $(3.5, 6)$ and $(-2.7, -6.4)$.

a Mark the two points on a graph.

b Draw the straight line that passes through the two points.

c Write the equation of the line.

d Draw the circle, centre $(0, 0)$, which passes through the two points.

e Write the equation of the circle.

Explore

◎ Draw a pair of axes, with both labelled from −10 to 10

◎ Choose a value for x between −10 and 10

◎ Follow the flowchart to produce two values for y (the positive and negative values)

◎ Follow the flowchart until you have finished

◎ Can you explain how it works?

◎ Flowcharts can be used whenever there is a simple set of instructions to follow; loci are produced by following simple steps

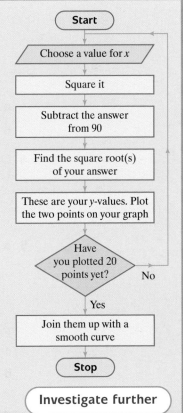

Loci

The following exercise tests your understanding of this chapter, with the questions appearing in order of increasing difficulty.

1 Draw any large triangle.
Construct the perpendicular bisector of each side.
The three bisectors should all meet at one point; label it C.
This point is called the 'circumcentre' of the triangle.
With compasses, using C as the centre, draw a circle that passes through all the vertices of the triangle.
Explain why this happens.

2 Draw any large triangle.
Construct the bisector of each angle.
The three bisectors should all meet at one point; label it I.
This point is called the 'incentre' of the triangle.
With compasses, using I as the centre, draw a circle that fits exactly inside the triangle – i.e. each side of the triangle is a tangent to this circle.
Explain why this happens.

3 A goat is tethered to the corner, A, of a 2 m by 2 m square post.
The post is fixed to the middle of a wall of a large grassy field.
The rope is 5 m long.

a Draw a scale diagram using the scale 1 cm to 1 m.

b Shade the area of grass on which the goat can feed.

4 On the same grid, draw the graphs of $x^2 + y^2 = 16$ and $y = 2x + 8$
Use your graphs to solve the simultaneous equations:

$$x^2 + y^2 = 16$$
$$y = 2x + 8$$

5 a i Draw two circles of radii 3 cm and 4 cm, with their centres 5 cm apart.

    ii At a point of intersection of the two circles, draw the tangent to each circle.

    iii Explain why the angle between these tangents is 90°.

For any pair of circles, if the distance between the centres is d, and the radii are r_1 and r_2, the circles are said to be 'orthogonal' if $d^2 = r_1^2 + r_2^2$

b Choose another Pythagorean triple and draw the related orthogonal circles.

OBJECTIVES

A*▶ **Examiners would normally expect students who get an A* grade to be able to:**

> Transform the graphs of $y = f(x)$, such as linear, quadratic, cubic, sine and cosine functions, using the transformations $y = f(x) + a$, $y = f(x + a)$, $y = af(x)$ and $y = f(ax)$

What you should already know ...

- Transform shapes using translations, reflections and enlargements
- Factorise quadratics including completing the square

- Draw, sketch and describe the graphs of linear, quadratic, sine and cosine functions

---

VOCABULARY

Translation – a transformation where every point moves the same distance in the same direction so that the object and the image are congruent

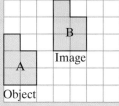

Shape A has been mapped onto Shape B by a translation of 3 units to the right and 2 units up

The vector for this would be $\begin{pmatrix} 3 \\ 2 \end{pmatrix}$

A translation is defined by the distance and the direction (vector)

Reflection – a transformation involving a mirror line (or axis of symmetry), in which the line from the shape to its image is perpendicular to the mirror line. To describe a reflection fully, you must describe the position or give the equation of its mirror line, for example, the triangle A is reflected in the mirror line $y = 1$ to give the image B

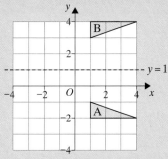

Stretch – a transformation that moves lines of an object parallel to another fixed line (usually the x-axis or y-axis)

Linear function – a function of the form $f(x) = ax + b$, where a and b are constants, and whose graph is a straight line with gradient a and y-intercept b; the equation of the graph of a linear function is usually expressed as $y = mx + c$, so m is the gradient and c is the y-intercept

Quadratic function – functions like $y = 3x^2$, $y = 9 - x^2$ and $y = 5x^2 + 2x - 4$ are quadratic functions; they include an x^2 term and may also include x terms and constants

The graphs of quadratic functions are always ∪-shaped or ∩-shaped

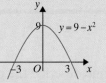

$y = ax^2 + bx + c$ is ∪-shaped when a is positive and ∩-shaped when a is negative

c is the intercept on the y-axis

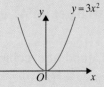

Note that other letters could be used as the variable instead of x (for example, $6t^2 - 3t - 5$ is also a quadratic expression and $h = 30t - 2t^2$ is a quadratic function)

Cubic function – functions of the form $f(x) = ax^3 + bx^2 + cx + d$ where a, b, c and d are constants and $a \neq 0$; the simplest cubic function is $f(x) = x^3$ (in this case, a is 1 and the other constants are zero)

The diagram shows the graph of $y = x^3$ and the graph of a more general cubic function, $y = x^3 - x^2 - 2x$

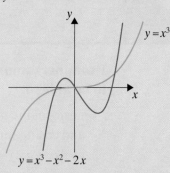

Sine (abbreviation sin) – in a right-angled triangle, the ratio of the length of the opposite side to the length of the hypotenuse

$$\sin x = \frac{opposite}{hypotenuse}$$

Cosine (abbreviation cos) – in a right-angled triangle, the ratio of the length of the adjacent side to the length of the hypotenuse

$$\cos x = \frac{adjacent}{hypotenuse}$$

Sketch – an outline of the graph with key features defined, for example, where the graph crosses the x-axis and y-axis

Learn 1   Transforming graphs of various functions

Examples: **a** This is the graph of $y = x^2$

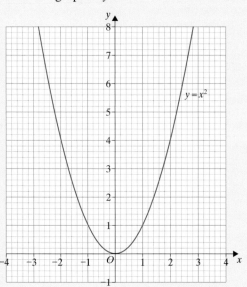

i Draw the graph of $y = x^2 + 2$
Each y-coordinate will be 2 more than its previous value.

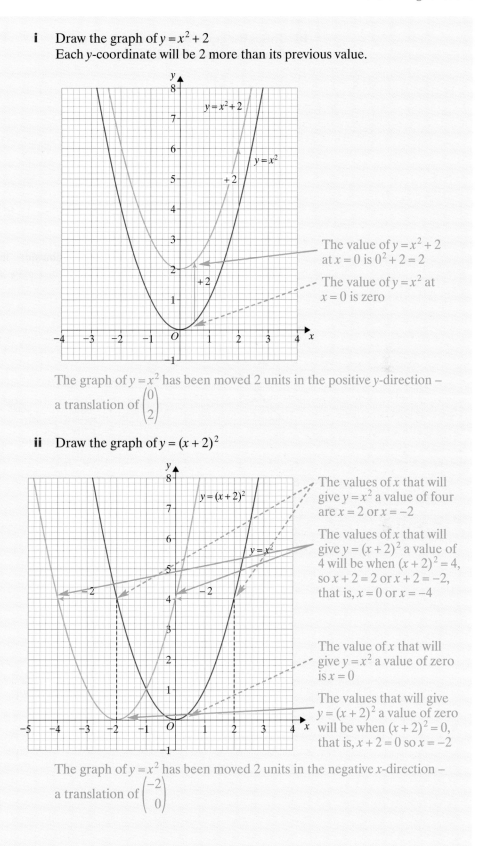

The value of $y = x^2 + 2$ at $x = 0$ is $0^2 + 2 = 2$

The value of $y = x^2$ at $x = 0$ is zero

The graph of $y = x^2$ has been moved 2 units in the positive y-direction – a translation of $\begin{pmatrix} 0 \\ 2 \end{pmatrix}$

ii Draw the graph of $y = (x + 2)^2$

The values of x that will give $y = x^2$ a value of four are $x = 2$ or $x = -2$

The values of x that will give $y = (x + 2)^2$ a value of 4 will be when $(x + 2)^2 = 4$, so $x + 2 = 2$ or $x + 2 = -2$, that is, $x = 0$ or $x = -4$

The value of x that will give $y = x^2$ a value of zero is $x = 0$

The values that will give $y = (x + 2)^2$ a value of zero will be when $(x + 2)^2 = 0$, that is, $x + 2 = 0$ so $x = -2$

The graph of $y = x^2$ has been moved 2 units in the negative x-direction – a translation of $\begin{pmatrix} -2 \\ 0 \end{pmatrix}$

iii This is the graph of the function $y = \sin x$.
Draw the graph of $y = 2\sin x$

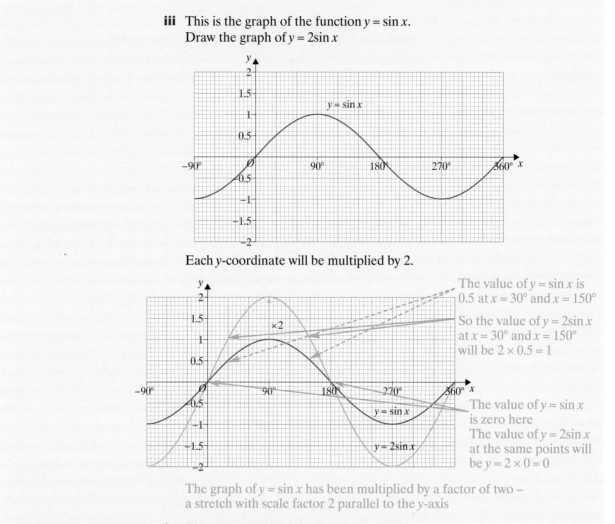

Each y-coordinate will be multiplied by 2.

The value of $y = \sin x$ is 0.5 at $x = 30°$ and $x = 150°$

So the value of $y = 2\sin x$ at $x = 30°$ and $x = 150°$ will be $2 \times 0.5 = 1$

The value of $y = \sin x$ is zero here
The value of $y = 2\sin x$ at the same points will be $y = 2 \times 0 = 0$

The graph of $y = \sin x$ has been multiplied by a factor of two – a stretch with scale factor 2 parallel to the y-axis

iv This is the graph of the function $y = \cos x$.
Draw the graph of $y = \cos 2x$

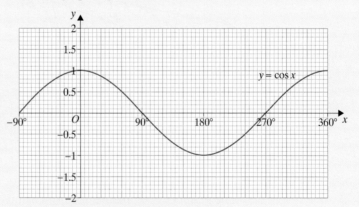

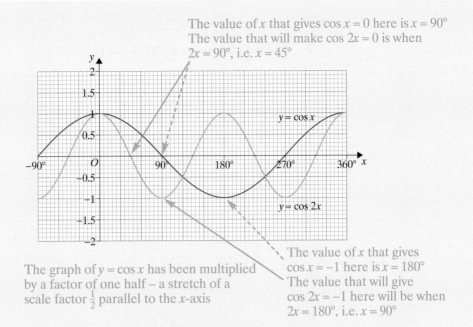

The value of x that gives $\cos x = 0$ here is $x = 90°$
The value that will make $\cos 2x = 0$ is when
$2x = 90°$, i.e. $x = 45°$

The graph of $y = \cos x$ has been multiplied by a factor of one half – a stretch of a scale factor $\frac{1}{2}$ parallel to the x-axis

The value of x that gives $\cos x = -1$ here is $x = 180°$
The value that will give $\cos 2x = -1$ here will be when $2x = 180°$, i.e. $x = 90°$

This table summarises the four transformations.

Graph	Description
$y = x^2 + 2$	Translation of the graph $y = x^2$ by the vector $\begin{pmatrix} 0 \\ 2 \end{pmatrix}$
$y = (x + 2)^2$	Translation of the graph $y = x^2$ by the vector $\begin{pmatrix} -2 \\ 0 \end{pmatrix}$
$y = 2\sin x$	The graph of $y = \sin x$ stretched by a scale factor 2 parallel to the y-axis
$y = \cos 2x$	The graph of $y = \cos x$ stretched by a scale factor $\frac{1}{2}$ parallel to the x-axis

In general, $y = f(x)$ is transformed as follows:

Graph	Description
$y = f(x) + a$	Translation $\begin{pmatrix} 0 \\ a \end{pmatrix}$
$y = f(x + a)$	Translation $\begin{pmatrix} -a \\ 0 \end{pmatrix}$
$y = af(x)$	Stretch by a scale factor a parallel to the y-axis
$y = f(ax)$	Stretch by a scale factor $\frac{1}{a}$ parallel to the x-axis

b This is the graph of $y = x^2 + 2x$

  i Translate the graph by the vector $\begin{pmatrix} 3 \\ 0 \end{pmatrix}$.

  ii Find the equation of the new graph in the form $y = ax^2 + bx + c$

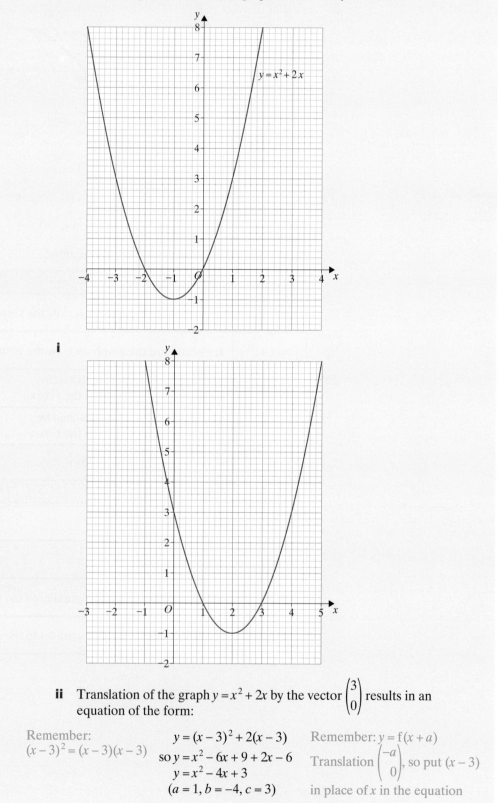

  i

  ii Translation of the graph $y = x^2 + 2x$ by the vector $\begin{pmatrix} 3 \\ 0 \end{pmatrix}$ results in an equation of the form:

Remember:
$(x - 3)^2 = (x - 3)(x - 3)$

$$y = (x - 3)^2 + 2(x - 3)$$
$$\text{so } y = x^2 - 6x + 9 + 2x - 6$$
$$y = x^2 - 4x + 3$$
$$(a = 1, b = -4, c = 3)$$

Remember: $y = f(x + a)$

Translation $\begin{pmatrix} -a \\ 0 \end{pmatrix}$, so put $(x - 3)$

in place of x in the equation

c By completing the square, or otherwise, find the transformation that maps the graph of $y = x^2$ onto the graph of:

i $y = x^2 + 10x + 25$

ii $y = x^2 + 6x + 10$

i $y = x^2 + 10x + 25 = (x + 5)^2$

The transformation is a translation of $\begin{pmatrix} -5 \\ 0 \end{pmatrix}$.

ii $y = x^2 + 6x + 10 = (x + 3)^2 + 10 - 9 = (x + 3)^2 + 1$

The transformation is a translation of $\begin{pmatrix} -3 \\ 1 \end{pmatrix}$.

Apply 1

1 a Sketch these graphs when translated by $\begin{pmatrix} 0 \\ 2 \end{pmatrix}$.

i $y = x$

ii $y = x^2$

iii $y = x^3$

iv $y = \sin x$

v $y = \cos x$

b Sketch these graphs when translated by $\begin{pmatrix} 3 \\ 0 \end{pmatrix}$.

i $y = x$

ii $y = x^2$

iii $y = x^3$

c Sketch these graphs when stretched by a scale factor of 3 parallel to the y-axis.

i $y = x$

ii $y = x^2$

iii $y = x^3$

iv $y = \sin x$

v $y = \cos x$

d Sketch these graphs when stretched by a scale factor of 2 parallel to the x-axis.

i $y = x$

ii $y = x^2$

iii $y = x^3$

iv $y = \sin x$

v $y = \cos x$

e Write the equation for all the transformed graphs in parts **a**–**d**.

2 For each set of graphs:

   a write the equation of the blue graph

   b by transforming the blue graph, write the equation of each red graph

   c write the transformation that produces each family of graphs.

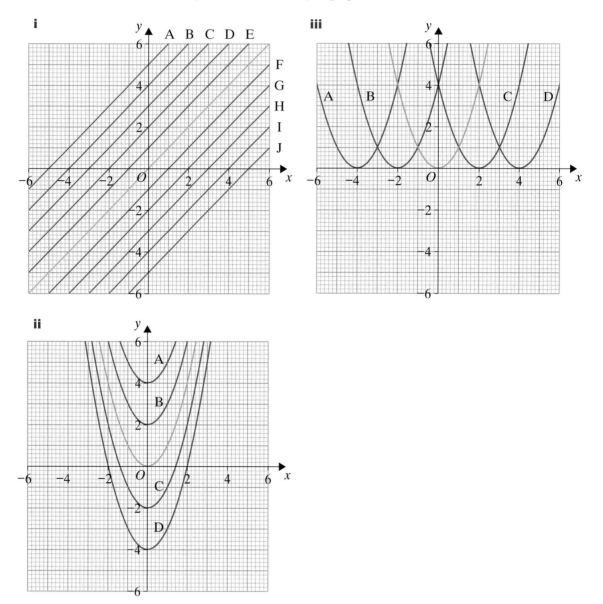

3 This is the graph of $y = x^3$

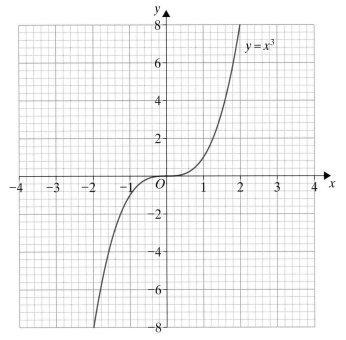

Using this graph, Faruq drew the graph of $y = (x - 2)^3$

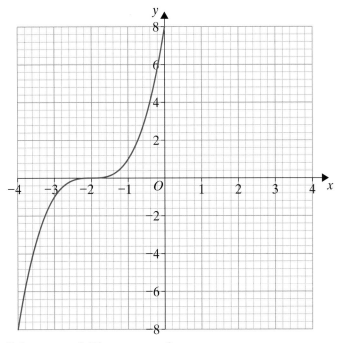

Is he correct? Give a reason for your answer.

4 For each set of graphs:

    a write the equation of the blue graph

    b by transforming the blue graph, write the equation of each graph

    c write the transformation that produces each family of graphs.

i

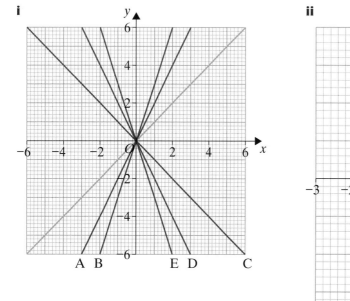

ii

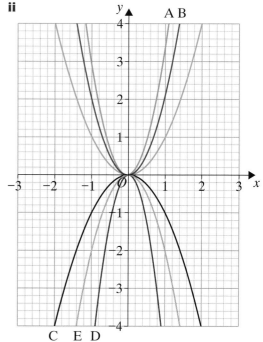

5 This is the graph of $y = \sin x$

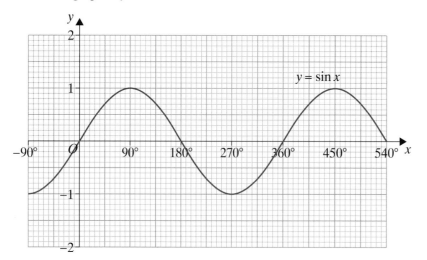

a Match each of these graphs with the correct equation.

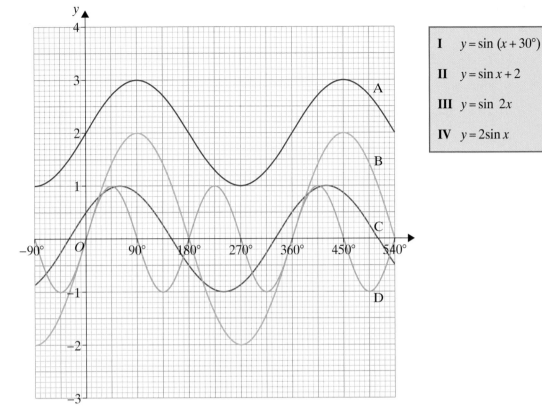

I	$y = \sin(x + 30°)$
II	$y = \sin x + 2$
III	$y = \sin 2x$
IV	$y = 2\sin x$

b Describe the transformation in each case.

6 Tiana is trying to draw the graph of $y = \cos 2x$
She draws the graph of $y = \cos x$ and applies a transformation.

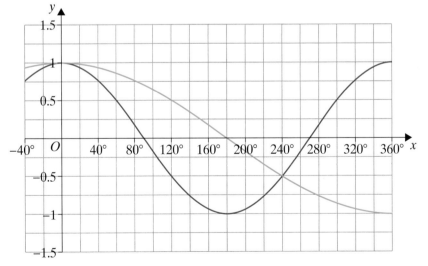

Has Tiana transformed $y = \cos x$ correctly?
Give a reason for your answer.

7 a Sketch the graph of $y = \cos x$ for $0 \leqslant x \leqslant 360°$ using the scale $1\ cm = 45°$ on the x-axis and $1\ cm = 1$ unit on the y-axis.

b Using this graph, sketch the graphs of:

   i $y = \cos x + 3$

   ii $y = 3\cos x$

   iii $y = \cos 3x$

   iv $y = \cos \dfrac{x}{3}$

8 This is the graph of $y = x^3 + 2x^2$

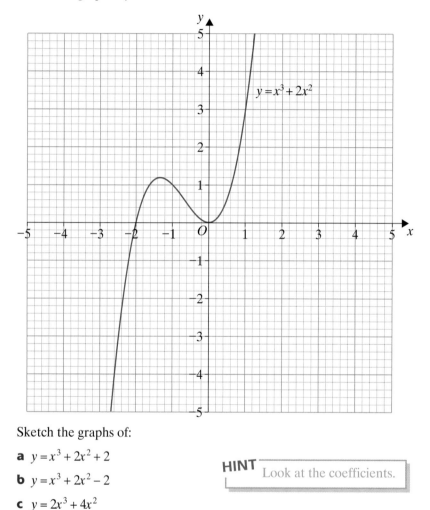

Sketch the graphs of:

a $y = x^3 + 2x^2 + 2$

b $y = x^3 + 2x^2 - 2$

c $y = 2x^3 + 4x^2$

> **HINT** Look at the coefficients.

9 Find two transformations that map the graph of $y = x^3$ onto the graph of $y = 8x^3$

10 This is the graph of $y = x^2 + 2x$

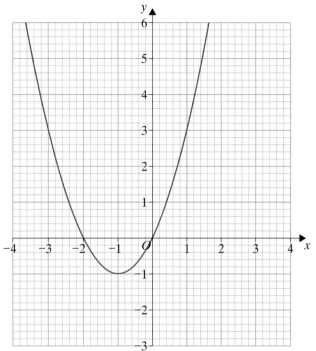

a i Translate the graph of $y = x^2 + 2x$ by $\begin{pmatrix} 0 \\ 2 \end{pmatrix}$

   ii Find the equation of the new graph in the form $y = ax^2 + bx + c$

   iii Check your answer using a graphic calculator or graphing software.

b i Translate the graph of $y = x^2 + 2x$ by $\begin{pmatrix} 0 \\ -3 \end{pmatrix}$

   ii Find the equation of the new graph in the form $y = ax^2 + bx + c$

   iii Check your answer using a graphic calculator or graphing software.

c i Translate the graph of $y = x^2 + 2x$ by $\begin{pmatrix} 4 \\ 0 \end{pmatrix}$

   ii Find the equation of the new graph in the form $y = ax^2 + bx + c$

   iii Check your answer using a graphic calculator or graphing software.

d i Translate the graph of $y = x^2 + 2x$ by $\begin{pmatrix} -3 \\ 0 \end{pmatrix}$

   ii Find the equation of the new graph in the form $y = ax^2 + bx + c$

   iii Check your answer using a graphic calculator or graphing software.

e i Translate the graph of $y = x^2 + 2x$ by $\begin{pmatrix} 1 \\ 2 \end{pmatrix}$

   ii Find the equation of the new graph in the form $y = ax^2 + bx + c$

   iii Check your answer using a graphic calculator or graphing software.

11 a Mary says it's easy to draw the graph of $y = x^2 + 4x + 4$

   Her method is to translate the graph of $y = x^2$ by $\begin{pmatrix} -2 \\ 0 \end{pmatrix}$.

   Explain why Mary's method works.

b Billy likes Mary's method. He realises he can draw the graph of

   $y = x^2 + 6x + 4$ by translating the graph of $y = x^2$ by $\begin{pmatrix} -3 \\ -5 \end{pmatrix}$.

   Explain why Billy's method works.

c Find the vector that translates $y = x^2$ onto the graph of $y = x^2 - 4x + 5$

d Find the vector that translates $y = x^2$ onto the graph of $y = x^2 + bx + c$

12 This is the graph of $y = 2\sin(x - 20)°$

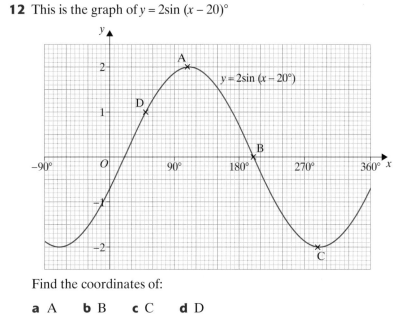

Find the coordinates of:

a A   **b** B   **c** C   **d** D

13 Get Real!

This graph shows the tidal pattern of 'The Havens'.

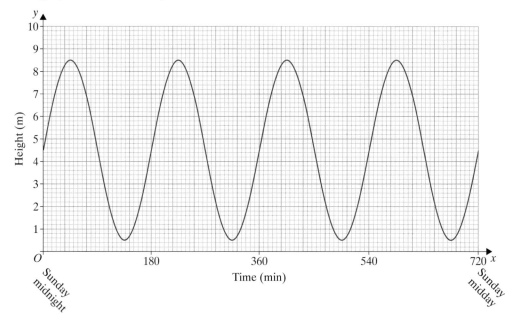

a The graph can be described by $y = a\sin(bx) + c$ by transforming the graph $y = \sin x$. Find a, b and c.

b The graph can also be created by transforming the graph $y = \cos x$. Find the equation of the graph using $y = \cos x$ as the starting point.

14 In each of these diagrams the equation of the red graph is $y = f(x)$
Find the equation of the transformed blue graph.

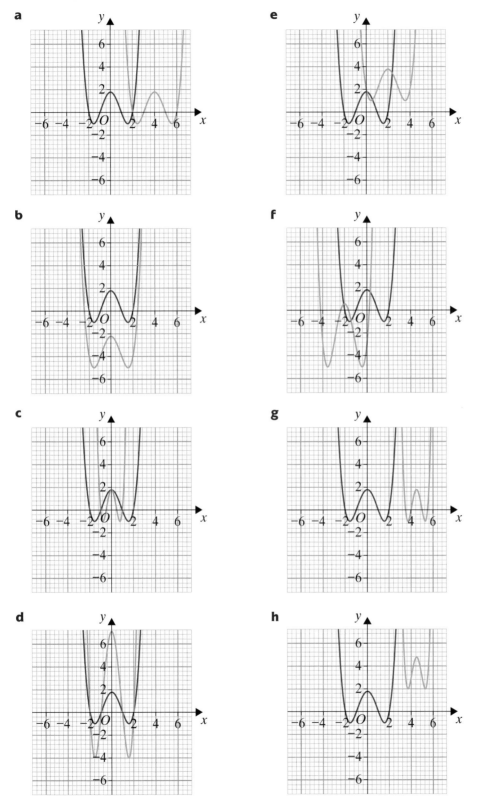

a

b

c

d

e

f

g

h

15 For each of these graphs, sketch:

   i $y = -f(x)$

   ii $y = f(-x)$

and describe the transformation in each case.

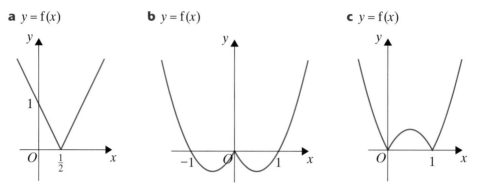

  a $y = f(x)$ **b** $y = f(x)$ **c** $y = f(x)$

Explore

◎ Draw accurately the graph of $y = \cos x$

◎ Copy and complete the table of values and plot the graph of $y = -\cos x$

x	0°	30°	45°	60°	90°	120°	135°	150°	180°	210°	225°	240°	270°	300°	315°	330°	360°
y	0																

◎ Copy and complete the table of values and plot the graph of $y = \cos(-x)$

x	0°	30°	45°	60°	90°	120°	135°	150°	180°	210°	225°	240°	270°	300°	315°	330°	360°
y	0																

                                                     Investigate further

Explore

◎ Find a function $y = f(x)$ such that $f(x) = f(-x)$

◎ Find a function $y = g(x)$ such that $g(-x) = -g(x)$

                                                     Investigate further

Explore

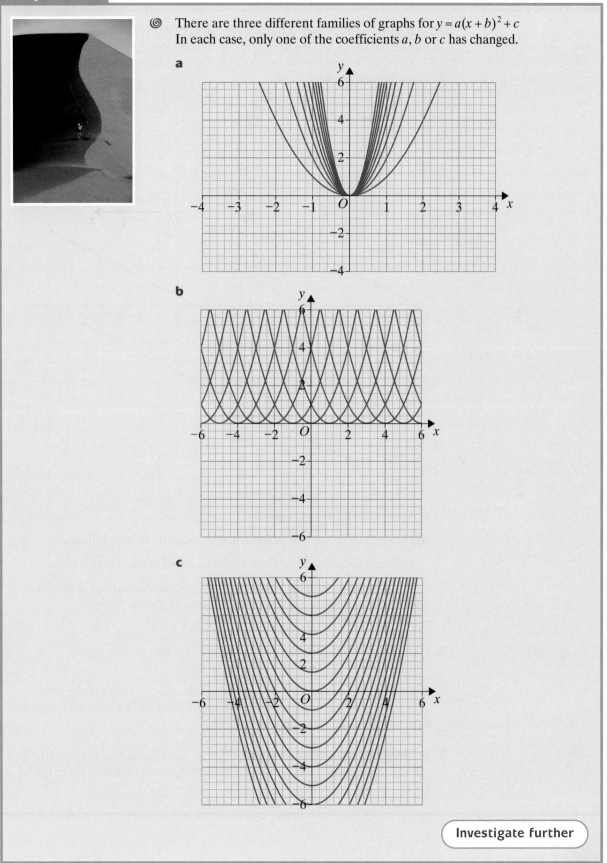

There are three different families of graphs for $y = a(x + b)^2 + c$
In each case, only one of the coefficients a, b or c has changed.

a

b

c

Investigate further

Transforming functions

The following exercise tests your understanding of this chapter, with the questions appearing in order of increasing difficulty.

1 Look at these graphs.

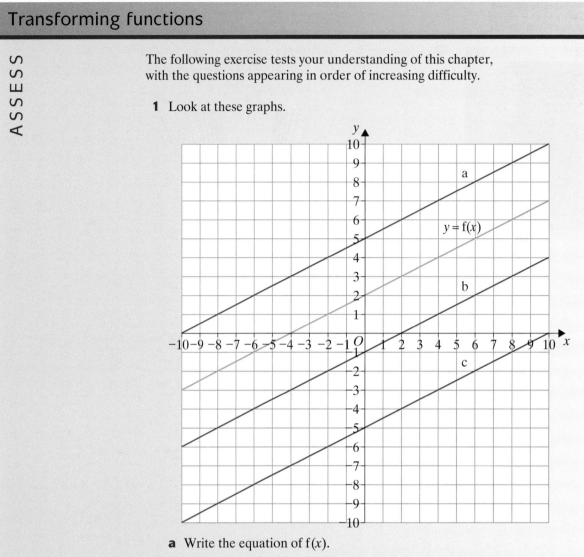

a Write the equation of $f(x)$.

b For each of the graphs labelled **a**, **b** and **c**, write in function terms:

   i their equations in the form $y = f(x) + a$ **and** in the form $y = f(x + a)$

   ii Use your answers to part **i** and work out their equations in the form $y = mx + c$ and show that you get the same answer for both transformations.

2 Draw on the same axes the graphs of $y = x^2$, $y = 2x^2$, $y = 3x^2$, $y = \frac{1}{2}x^2$ and $y = \frac{1}{4}x^2$

Writing the equation of $y = x^2$ as $y = f(x)$ write the equations of the other four graphs as functional equations. Explain what the form $y = af(x)$ means when compared to the equation $y = f(x)$

3 Starting with a sketch of the graph of $y = x^2$, use the method of completing the square to show how you can use transformations to sketch the graph of $y = x^2 - 6x + 8$

20 Algebraic proofs

OBJECTIVES

D **Examiners would normally expect students who get a D grade to be able to:**

Decide with a reason whether a harder statement is true or false

Identify a counter example

C **Examiners would normally expect students who get a C grade also to be able to:**

Understand the difference between a demonstration and a proof

Show step-by-step deductions in providing a basic algebraic explanation

B **Examiners would normally expect students who get a B grade also to be able to:**

Show step-by-step deductions in providing a full mathematical explanation

A **Examiners would normally expect students who get an A grade also to be able to:**

Derive simple algebraic proofs using reasoning

A* **Examiners would normally expect students who get an A* grade also to be able to:**

Derive harder algebraic proofs using reasoning and logic

What you should already know ...

- Odd and even numbers
- Prime numbers
- Factors and multiples
- Square and cube numbers

VOCABULARY

Sum – the result of adding together two (or more) numbers, variables, terms or expressions

Product – the result of multiplying together two (or more) numbers, variables, terms or expressions

Consecutive numbers – whole numbers that are next to each other on the number line in sequence

Counter example – an example that disproves a statement. For example, the statement 'All prime numbers are odd', is not true because 2 is a prime number and 2 is not odd – this is a counter example

Learn 1 Algebraic proofs

Examples:

a If R is an odd number and S is an even number, state whether each of the following is odd or even and give a reason for your answer.

i $R + S$
ii $R \times S$

i $R + S$ is odd because an odd number + an even number is always odd

Try out different values of R and S to show the result

In general: $O + O = E$
$O + E = O$
$E + O = O$
$E + E = E$

ii $R \times S$ is even because an odd number × an even number is always even

Try out different values of R and S to show the result

In general: $O \times O = O$
$O \times E = E$
$E \times O = E$
$E \times E = E$

b Trevor says that when you square a number, the answer is always greater than or equal to the original number.
Give an example to show that Trevor is wrong.

Try positive then negative then decimal (fraction) numbers.

Try 1 $1^2 = 1$ so true when the number = 1
Try 2 $2^2 = 4$ so true when the number = 2
Try 3 $3^2 = 9$ so true when the number = 3
Try –1 $(-1)^2 = 1$ so true when the number = –1
Try –2 $(-2)^2 = 4$ so true when the number = –2
Try 0.5 $0.5^2 = 0.25$ so **not** true when the number = 0.5

Sometimes it will be necessary to try out a few examples to find the solution

Trevor is wrong as $0.5^2 = 0.25$ which is smaller than the original number.

Giving a numerical example that disproves a statement is often called a counter example

Numbers such as 0 and 1 as well as prime numbers, negative numbers and fractional numbers are all useful examples to try out when trying to find a counter example in questions like this

c Prove that if n is a positive integer then $n^2 - n$ is an even number.

You might find it helpful to start by checking that if n is a positive integer then $n^2 - n$ is an even number for different values of n

In examples like this one it is often helpful to consider the two cases when n is an even number and when n is an odd number

If n is an even number then it can be written as $2a$.

Remember that $2a$ is always an even number

$n^2 - n = (2a)^2 - 2a = 4a^2 - 2a = 2(2a^2 - a)$

$2(2a^2 - a)$ is also an even number

since any whole number multiplied by 2 is an even number.

If n is an odd number then it can be written as $2a + 1$

Remember that $2a + 1$ is always an odd number

$n^2 - n = (2a + 1)^2 - (2a + 1) = (4a^2 + 4a + 1) - (2a + 1) = 4a^2 + 2a = 2(2a^2 + 2a)$

$2(2a^2 + 2a)$ is an even number

since any whole number multiplied by 2 is an even number

So if n is a positive integer then $n^2 - n$ is an even number.

Alternatively:

If n is an even number then

n^2 is an even number squared, which is an even number

$n^2 - n$ is an even − even = even

If n is an odd number then

n^2 is an odd number squared, which is an odd number

$n^2 - n$ is an odd − odd = even

d Prove that the difference between two consecutive square numbers is always an odd number.

Let the first number be x and the square of this number is x^2

The next consecutive number is $x + 1$ and the square of this number is $(x + 1)^2$

We need to prove that the difference between x^2 and $(x + 1)^2$ is an odd number.

The difference can be written as
$$(x + 1)^2 - x^2$$
$$= (x^2 + 2x + 1) - x^2$$
$$= 2x + 1$$

It is a good idea to take the smallest number from the largest number when finding the difference ... so that you don't get a negative answer

Since the number $2x$ is always an even number then the number $2x + 1$ must be an odd number so that the difference between two consecutive square numbers is always an odd number.

Apply 1

1 If n is a positive integer:

   a explain why $2n$ must be an even number

   b explain why $2n + 1$ must be an odd number

   c explain why $n(n + 1)$ is always an even number.

2 Is the value of $3n - 1$ always an even number?
Give a reason for your answer.

3 Is the value of $5n + 1$ always an even number?
Give a reason for your answer.

4 If p is an odd number explain why $p^2 + 1$ is an even number.

5 If p is an odd number and q is an even number explain why $(p - q)(p + q)$ is an odd number.

6 P is a prime number.
Q is an odd number.

State whether each of the following is:

i always even

ii always odd

iii could be either even or odd.

   a PQ

   b $P(Q + 1)$

7 k is an odd number.
Gail says that $\frac{1}{2}k + \frac{1}{2}$ is always even.
Give a counter example to show that Gail is wrong.

8 Christine says that all numbers have an even number of factors.
Give a counter example to show that Christine is wrong.

9 The product of two consecutive numbers is divisible by 12.
Write down two numbers that satisfy this statement.

10 The product of two consecutive numbers is divisible by 20.
Write down two numbers that satisfy this statement.

11 Duncan says that the square of a number is always bigger than the
number itself.
Is Duncan correct?
Give a reason for your answer.

12 Prove that the difference between any two even numbers is always an
even number.

13 Prove that the sum of three consecutive numbers is equal to three times
the middle number.

14 Prove that the sum of two consecutive odd numbers is always a multiple of 4.

15 Paul says that an odd number squared is always an odd number.
Is Paul correct?
Give a reason for your answer.

16 John says that $(a + b)^2 = a^2 + b^2$
Is John correct?
Give a reason for your answer.

17 Mick says that when you square a number the answer is always greater than zero.
Is Mick correct?
Give a reason for your answer.

18 Keith says that the cube of a number is always bigger than the square of that number.
Is Keith correct?
Give a reason for your answer.

19 Part of a number grid is shown below:

1	2	3	4	5	6	7	8
9	10	11	12	13	14	15	16
17	18	19	20	21	22	23	24
25	26	27	28	29	30	31	32
33	34	35	36	37	38	39	40
41	42	43	44	45	46	47	48
49	50	51	52	53	54	55	56
57	58	59	60	61	62	63	64

The shaded shape is called T_{12} because it has 12 in the middle of the top row.
The sum of the numbers in T_{12} is 56.

a This is T_n

Copy and fill in the empty boxes on T_n

b Find the sum of all the numbers in T_n in terms of n.
Give your answer in its simplest form.

c Explain why the sum of all the numbers in T_n is always divisible by 4.

20 Part of a number grid is shown below:

1	2	3	4	5	6	7	8
9	10	11	12	13	14	15	16
17	18	19	20	21	22	23	24
25	26	27	28	29	30	31	32
33	34	35	36	37	38	39	40
41	42	43	44	45	46	47	48
49	50	51	52	53	54	55	56
57	58	59	60	61	62	63	64

The shaded shape is called B_{11} because it has 11 in the top left-hand corner.

a This is B_n

Copy and fill in the empty boxes on B_n

b Amara notices that $\quad 11 \times 20 = 220$
and that $\quad\quad\quad\quad 19 \times 12 = 228$

Show, using algebra, that the difference of the products of the diagonals is always 8.

21 Prove that the product of an even number and an odd number is always an even number.

22 Prove that the product of three consecutive even numbers is a multiple of 8.

23 Prove that, for any positive integer n, $n^2 + n$ is divisible by 2.

24 The nth term of a triangle number is given by the formula $\frac{1}{2}n(n+1)$
Prove that the sum of any two consecutive triangle numbers is always a square number.

Explore

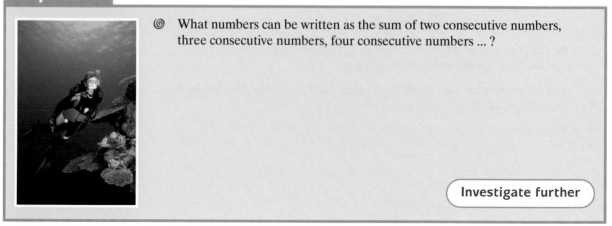

◎ What numbers can be written as the sum of two consecutive numbers, three consecutive numbers, four consecutive numbers ... ?

Investigate further

Explore

⊚ A quick way to add the numbers from 1 to 100 is shown below:

Write out the numbers $1 + 2 + 3 + 4 + ... + 97 + 98 + 99 + 100$
Reverse the numbers $100 + 99 + 98 + 97 + ... + 4 + 3 + 2 + 1$

Adding the two series $101 + 101 + 101 + 101 + ... + 101 + 101 + 101 + 101$

Each pair of numbers adds up to 101 and there are 100 pairs of numbers
So the sum of all the numbers written down is 100×101 (that is, 100 lots of the total of 101)

However this is twice the total so

$$1 + 2 + 3 + 4 + ... + 97 + 98 + 99 + 100 = \tfrac{1}{2} \times 100 \times 101$$
$$= \tfrac{1}{2} \times 10100$$
$$= 5050$$

⊚ Can you add the numbers from 1 to 1000 using this method?

⊚ Can you add the numbers from 1 to n (where n is any number) using this method?

(Investigate further)

Algebraic proofs

ASSESS

The following exercise tests your understanding of this chapter, with the questions appearing in order of increasing difficulty.

1 Ross says that the product of two prime numbers is always odd.
Give a counter example to show that Ross is wrong.

2 P is a prime number.
Q is an even number.

State whether each of the following is:

i always even

ii always odd

iii could be either even or odd.

a PQ

b $P(Q - 1)$

3 Prove that the sum of four consecutive numbers is always an even number.

4 Prove that the sum of three consecutive even numbers is always a multiple of six.

5 Zoë says that when you double a prime number and add one, the answer is another prime.
Is Zoë correct?
Give a reason for your answer.

6 Prove that the mean of three consecutive numbers is equal to the middle number.

7 Part of a number grid is shown below:

1	2	3	4	5	6	7
8	9	10	11	12	13	14
15	16	17	18	19	20	21
22	23	24	25	26	27	28
29	30	31	32	33	34	35
36	37	38	39	40	41	42
43	44	45	46	47	48	49

The shaded shape is called L_{16} because it has 16 in the corner.
The sum of the numbers in L_{16} is 60.

a This is L_n

Copy and fill in the empty boxes on L_n

b Find the sum of all the numbers in L_n in terms of n.
Give your answer in its simplest form.

c Explain why the sum of all the numbers in L_n is always divisible by 4.

8 Prove that when you take the product of any two consecutive odd numbers and then add 1, the answer is always a multiple of 4.

9 Prove that the sum of the squares of any two odd numbers will leave a remainder of 2 when divided by 4.

Try some real past exam questions to test your knowledge:

10 a n is a positive integer.

    i Explain why $n(n + 1)$ must be an even number.

    ii Explain why $2n + 1$ must be an odd number.

b Expand and simplify $(2n + 1)^2$

c Prove that the square of any odd number is always 1 more than a multiple of 8.

Spec A, Higher Paper 1, June 2004

11 The **two-digit** number 37 can be written as $(10 \times 3) + 7$
Similarly the **two-digit** number ab can be written as $10a + b$

a def is a three-digit number.
Write an expression for the number def in terms of d, e and f.

b Write an expression for the three-digit number fed in terms of f, e and d.

c Using your answers for parts **a** and **b**, write down and simplify an expression for $def - fed$.

d Hence show that $def - fed$ is divisible by 9.

Spec A, Higher Paper 1, Nov 2005

Glossary

Acceleration – the rate of change of velocity with respect to time

Adjacent side – in a right-angled triangle, the shorter side adjacent to the known angle

Adjacent

Alternate angles – the angles marked *a*, which appear on opposite sides of the transversal

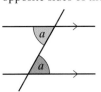

Angle bisector – a line that divides an angle into two equal parts

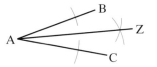

AZ is the angle bisector of angle BAC

Angle of depression – the angle below the horizontal between a line and the horizontal. Example: in the diagram, the angle *a* is the angle of depression of point B from point A

Angle of elevation – the angle above the horizontal between a line and the horizontal. Example: in the diagram, the angle *b* is the angle of elevation of point A on top of the cliff from point B

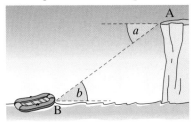

Arc (of a circle) – part of the circumference of a circle; a minor arc is less than half the circumference and a major arc is greater than half the circumference

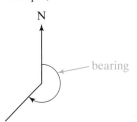

Minor sector   Minor arc

Major sector

Major arc

Bearing – an angle measured clockwise from North; all bearings should be written as three figure numbers, for example, 125° or 045°

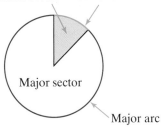

N

bearing

Bisect – to divide into two equal parts

Brackets – these show that the terms inside should be treated alike, for example,

$2(3x + 5) = 2 \times 3x + 2 \times 5 = 6x + 10$

Certain – an outcome with probability 1, for example, the sun rising and setting

Chord – a straight line joining two points on the circumference of a circle

Circumference – the perimeter of a circle

Coefficient – the number (with its sign) in front of the letter representing the unknown, for example:

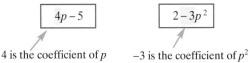

$4p - 5$ $2 - 3p^2$

4 is the coefficient of *p*   −3 is the coefficient of p^2

Collinear – points lying in a straight line

Completing the square – this refers to writing a quadratic expression in the form $(x + a)^2 + b$ where a and b are positive or negative constants (that is, writing it as a squared term and a constant)

This is useful in solving equations and finding the maximum or minimum value of a quadratic expression

Compound measure – a measure formed from two or more other measures, for example,

$$\text{speed} \left(= \frac{\text{distance}}{\text{time}}\right), \text{density} \left(= \frac{\text{mass}}{\text{volume}}\right),$$

$$\text{population density} \left(= \frac{\text{population}}{\text{area}}\right)$$

Cone – a pyramid with a circular base and a curved surface rising to a vertex

Congruent – exactly the same size and shape; one of the shapes might be rotated or flipped over

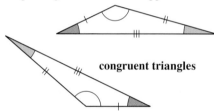

congruent triangles

Consecutive numbers – whole numbers that are next to each other on the number line in sequence

Converse of Pythagoras' theorem – in any triangle, if $c^2 = a^2 + b^2$ then the triangle has a right angle opposite c; for example, if $c = 17$ cm, $a = 8$ cm, $b = 15$ cm, then $c^2 = a^2 + b^2$ so this is a right angle

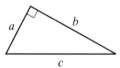

Conversion factor – the number by which you multiply or divide to change measurements from one unit to another. The approximate conversion factors that you should know are:

Length	Mass	Capacity
1 foot ≈ 30 cm	1 kg ≈ 2.2 pounds	1 gallon ≈ 4.5 litres
5 miles ≈ 8 km		1 litre ≈ 1.75 pints

The table below gives conversion factors for metric units of length, mass and capacity.

Metric system

Length	Mass	Capacity
1 cm = 10 mm	1 g = 1000 mg	1 ℓ = 100 cℓ or 1000 mℓ
1 m = 100 cm or 1000 mm	1 kg = 1000 g	
1 km = 1000 m	1 t = 1000 kg	

Corresponding angles – the angles marked c, which appear on the same side of the transversal

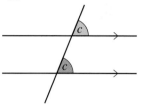

Cosine (abbreviation cos) – in a right-angled triangle, the ratio of the length of the adjacent side to the length of the hypotenuse

$$\cos x = \frac{adjacent}{hypotenuse}$$

Counter example – an example that disproves a statement. For example, the statement 'All prime numbers are odd', is not true because 2 is a prime number and 2 is not odd – this is a counter example

Cube – a solid with six identical square faces

Cube number – a cube number is the outcome when a whole number is multiplied by itself then multiplied by itself again; cube numbers are 1, 8, 27, 64, 125, ...

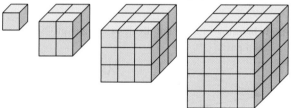

The rule for the nth term of the cube numbers is n^3

Cube root – the cube root of a number such as 125 is a number whose outcome is 125 when multiplied by itself then multiplied by itself again

Cubic function – functions of the form $f(x) = ax^3 + bx^2 + cx + d$ where a, b, c and d are constants and $a \neq 0$; the simplest cubic function is $f(x) = x^3$ (in this case, a is 1 and the other constants are zero)

The diagram shows the graph of $y = x^3$ and the graph of a more general cubic function, $y = x^3 - x^2 - 2x$

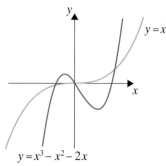

Cuboid – a solid with six rectangular faces (two or four of the faces can be squares)

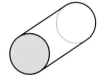

Cyclic quadrilateral – a quadrilateral whose vertices lie on the circumference of a circle

Cylinder – a prism with a circle as a cross-sectional face

Decimal places – the digits to the right of a decimal point in a number, for example, in the number 23.657, the number 6 is the first decimal place (worth $\frac{6}{10}$), the number 5 is the second decimal place (worth $\frac{5}{100}$), and 7 is the third decimal place (worth $\frac{7}{1000}$); the number 23.657 has 3 decimal places

Density – to calculate density, divide the mass of the object by the volume of the object. It is usually given in grams per cubic centimetre (g/cm^3) or kilograms per cubic metre (kg/m^3)

In the triangle, cover the item you want, then the rest tells you what to do

$$\text{Density} = \frac{\text{mass}}{\text{volume}}$$

$\frac{\text{kilograms}}{\text{cubic metres}}$ gives kg/m^3

$$\text{Mass} = \text{density} \times \text{volume}$$

$$\text{Volume} = \frac{\text{mass}}{\text{density}}$$

Dependent events – events are dependent when the outcome of one affects the outcome of the other, for example, taking two successive balls from a box without replacing the first one

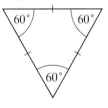

Diameter – a chord passing through the centre of a circle; the diameter is twice the length of the radius

Discriminant – the expression $b^2 - 4ac$, which is part of the quadratic formula

Edge – a line segment that joins two vertices of a solid

Enlargement – an enlargement changes the size of an object (unless the scale factor is 1) but not its shape; it is defined by giving the centre of enlargement and the scale factor; the object and the image are similar

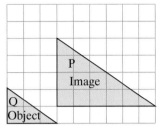

Triangle P is an enlargement of triangle Q
All the lines have doubled in size
The scale factor of the enlargement is 2

Equation – a statement showing that two expressions are equal, for example, $2y - 7 = 15$

Equidistant – the same distance; if A is equidistant from B and C, then AB and AC are the same length

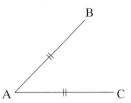

Equilateral triangle – a triangle with 3 equal sides and 3 equal angles – each angle is 60°

Evens – probability $\frac{1}{2}$, for example, there is an even chance of getting a head or a tail when you toss a coin

Experimental probability or **relative frequency** – this is found by experiment, for example, if you get 6 heads and 4 tails, the experimental probability would be $\frac{6}{10}$ or 0.6 for getting a head

Exponential decay – similar to exponential growth, but the amount decreases instead of increasing; the rate of decay is proportional to the amount remaining

The diagram shows graphs of exponential growth and decay; note the similarities and differences

Exponential growth and decay

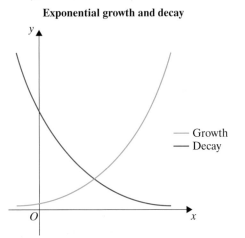

— Growth
— Decay

Exponential functions – functions of the form $y = Ak^{mx}$ where A, k and m are constants, for example, if £1500 is invested at 2.5% compound interest, the amount of money, £y, in the account after x years is given by the formula $y = 1500 \times 1.025^x$, which is an exponential function with $A = 1500$, $k = 1.025$ and $m = 1$

The amount of money in the account is said to **grow exponentially**, which does not necessarily mean that it grows very rapidly, but that its rate of growth is proportional to the amount in the account

Expression – a mathematical statement written in symbols, for example, $3x + 1$ or $x^2 + 2x$

Face – one of the flat surfaces of a solid

Formula – an equation showing the relationship between two or more variables, for example, $E = mc^2$

Front elevation – a diagram of a 3-D solid showing the view from the front, for example,

Front elevation

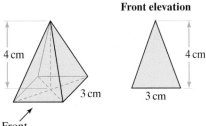

4 cm

3 cm

4 cm

3 cm

Front

In some cases, as in this prism, the elevation from the front is the same as its cross-section

They are congruent trapezia

Cross-section

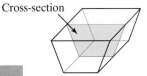

Gradient – a measure of how steep a line is

$$\text{Gradient} = \frac{\text{change in vertical distance}}{\text{change in horizontal distance}} = \frac{y}{x}$$

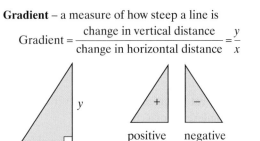

y

x

+ positive gradient

− negative gradient

Hemisphere – a half sphere

Hypotenuse – the longest side of a right-angled triangle, opposite the right angle

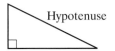

Hypotenuse

Identity – two expressions linked by the $\equiv$ sign are true for all values of the variable, for example, $3x + 3 \equiv 3(x + 1)$

Image – the shape after it undergoes a transformation, for example, reflection, rotation, translation or enlargement

Imperial units – these are units of measurement historically used in the United Kingdom and other English-speaking countries; they are now largely replaced by metric units. Imperial units include:

- inches (in), feet (ft), yards (yd) and miles for lengths
- ounces (oz), pounds (lb), stones and tons for mass
- pints (pt) and gallons (gal) for capacity

Impossible – an outcome with a probability 0, for example, the sun turning green

Independent events – events are independent when the outcome of one does not affect the outcome of the other, for example, tossing a coin and drawing a card from a pack

Inequality – statements such as $a \neq b$, $a \leqslant b$ or $a > b$ are inequalities

Inequality signs – $<$ means less than, $\leqslant$ means less than or equal to, $>$ means greater than, $\geqslant$ means greater than or equal to

Integer – any positive or negative whole number or zero, for example, -2, -1, 0, 1, 2 ...

Intercept – the *y*-coordinate of the point at which the line crosses the *y*-axis

Isosceles triangle – a triangle with 2 equal sides and 2 equal angles; the equal angles are called **base angles**

Kite – a quadrilateral with two pairs of equal adjacent sides

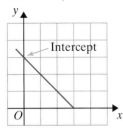

Likely – an outcome with a probability greater than $\frac{1}{2}$, for example, rain falling in November in the UK

Linear equation – an equation where the highest power of the variable is 1; for example, $3x + 2 = 7$ is a linear equation but $3x^2 + 2 = 7$ is not

Linear function – a function of the form $f(x) = ax + b$, where *a* and *b* are constants, and whose graph is a straight line with gradient *a* and *y*-intercept *b*; the equation of the graph of a linear function is usually expressed as $y = mx + c$, so *m* is the gradient and *c* is the *y*-intercept

Linear graph – the graph of a linear function of the form $y = mx + c$; if *c* is zero, the graph is a straight line through the origin (the point $(0, 0)$) indicating that *y* is directly proportional to *x*; if *m* is zero, the graph is parallel to the *x*-axis

Loci – the plural of locus

Locus – the path followed by a moving point; for example, the locus of a point, A, keeping at a fixed distance of 5 cm from a point B, is a circle, centre B, radius 5 cm, or, in 3 dimensions, a sphere, centre B, radius 5 cm

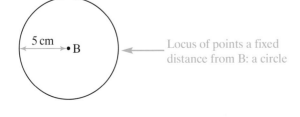

Locus of points a fixed distance from B: a circle

Lower bound – this is the minimum possible value of a measurement, for example, if a length is measured as 37 cm correct to the nearest centimetre, the lower bound of the length is 36.5 cm

Mapping – a transformation or enlargement is often referred to as a mapping with points on the object mapped onto points on the image

Metric units – these are related by multiples of 10 and include:
- metres (m), millimetres (mm), centimetres (cm) and kilometres (km) for lengths
- grams (g), milligrams (mg), kilograms (kg) and tonnes (t) for mass
- litres (ℓ), millilitres (mℓ) and centilitres (cℓ) for capacity

Midpoint – the middle point of a line

Mutually exclusive events – these cannot both happen in the same experiment, for example, getting a head and a tail on one toss of a coin

Net – a two-dimensional shape made of polygons that can be folded to make a three-dimensional solid, for example,

Net of a cuboid   Net of a triangular prism

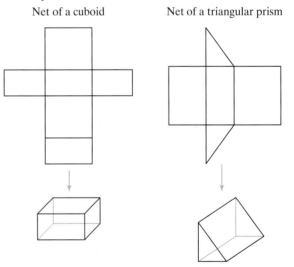

Non-linear equation – an equation that cannot be represented by a straight line graph, for example, $3x^2 + 2 = 7$, $x^2 + y^2 = 25$ and $xy = 10$ are non-linear equations

Number line – a line where numbers are represented by points upon it; simple inequalities can be shown on a number line

Object – the shape before it undergoes a transformation, for example, translation or enlargement

Opposite angles – the angles marked *a*, which are formed when two line segments intersect

Opposite side – in a right-angled triangle, the side opposite the known angle

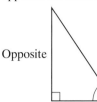

Opposite

Outcome – the result of an experiment, for example, when you toss a coin, the outcome is a head or a tail

Parallel lines – two lines that never meet and are always the same distance apart

Parallelogram – a quadrilateral with opposite sides equal and parallel

Perpendicular – at right angles to; two lines at right angles to each other are perpendicular lines

Perpendicular bisector – a line at right angles to a given line that also divides the given line into two equal parts

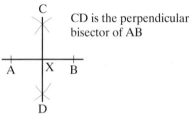

CD is the perpendicular bisector of AB

Perpendicular lines – two lines at right angles to each other

Plan – a diagram of a 3-D solid showing the view from above; these diagrams show a square-based pyramid and its plan

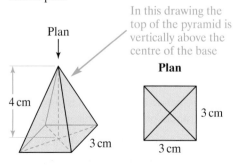

Plan

In this drawing the top of the pyramid is vertically above the centre of the base

Plan

4 cm

3 cm

3 cm

3 cm

3 cm

Prism – a three-dimensional solid with two cross-sectional faces that are identical polygons, parallel to each other; all other faces are either parallelograms or rectangles

Prisms are named according to the cross-sectional face; for example,

Triangular prism   Hexagonal prism   Parallelogram prism

Probability – a value between 0 and 1 (which can be expressed as a fraction, decimal or percentage) that gives the likelihood of an event

Product – the result of multiplying together two (or more) numbers, variables, terms or expressions

Proof – a series of logical mathematical steps that confirms the truth of a mathematical statement

Prove – to provide a proof

Pyramid – a solid with a polygon as the base and one other vertex; all the vertices of the base are joined to this vertex forming triangular faces. Pyramids are named according to their base, for example,

Square pyramid   Triangular pyramid

Pythagoras' theorem – in a right-angled triangle, the square of the length of the hypotenuse is equal to the sum of the squares of the lengths of the other two sides

$$c^2 = a^2 + b^2$$

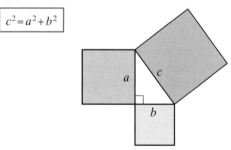

The area of the largest square = the total area of the two smaller squares

Pythagorean triple – a set of three integers a, b, c that satisfies $c^2 = a^2 + b^2$; for example, 3, 4, 5 ($5^2 = 3^2 + 4^2$), 5, 12, 13 ($13^2 = 5^2 + 12^2$), 6, 8, 10 ($10^2 = 6^2 + 8^2$) and 15, 36, 39 ($39^2 = 15^2 + 36^2$)

Quadratic equation – an equation that includes an x^2 term and may also include x terms and constants, for example, $5x^2 + 2x - 4 = 0$; in this example the **coefficient** of x^2 is 5, the coefficient of x is 2 and -4 is the constant term

Solutions of a quadratic equation can be found by graphical methods or (more accurately) by factorising, using the formula or completing the square

Quadratic expression – an expression containing terms where the highest power of the variable is 2

Quadratic expressions	Non-quadratic expressions
x^2	x
$x^2 + 2$	$2x$
$3x^2 + 2$	$\frac{1}{x}$
$4 + 4y^2$	$3x^2 + 5x^3$
$(x + 1)(x + 2)$	$x(x + 1)(x + 2)$

Quadratic formula – the solutions (sometimes called the 'roots') of the quadratic equation $ax^2 + bx + c = 0$ are given by the quadratic formula:

$$x = \frac{-b \pm \sqrt{b^2 - 4ac}}{2a}$$

These solutions are the x-coordinates of the points of intersection of $y = ax^2 + bx + c$ with the x-axis

Quadratic function – functions like $y = 3x^2$, $y = 9 - x^2$ and $y = 5x^2 + 2x - 4$ are quadratic functions; they include an x^2 term and may also include x terms and constants

The graphs of quadratic functions are always $\cup$-shaped or $\cap$-shaped

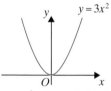

$y = ax^2 + bx + c$ is $\cup$-shaped when a is positive and $\cap$-shaped when a is negative

c is the intercept on the y-axis

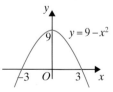

Note that other letters could be used as the variable instead of x (for example, $6t^2 - 3t - 5$ is also a quadratic expression and $h = 30t - 2t^2$ is a quadratic function)

Radius – the distance from the centre of a circle to any point on the circumference

Random – a choice made when all outcomes are equally likely, for example, picking a raffle ticket from a box with your eyes shut

Ratio – the ratio of two or more numbers or quantities is a way of comparing their sizes, for example, if a school has 25 teachers and 500 students, the ratio of teachers to students is 25 to 500, or 25 : 500 (read as 25 to 500)

Real number – any rational or irrational number; all real numbers can be represented on the number line

$$-5 \quad 42 \quad 0 \quad \pi \quad \sqrt{2} \quad \tfrac{4}{9}$$

Reciprocal functions – functions which involve negative powers of the variable; the simplest reciprocal function is $y = x^{-1}$, more commonly written as $y = \frac{1}{x}$ or $xy = 1$

This function does not exist for $x = 0$ so its graph has two separate parts, which approach the axes very closely but never meet them

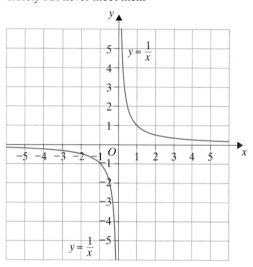

Reciprocal functions can represent relationships of inverse proportion

Reflection – a transformation involving a mirror line (or axis of symmetry), in which the line from the shape to its image is perpendicular to the mirror line. To describe a reflection fully, you must describe the position or give the equation of its mirror line, for example, the triangle A is reflected in the mirror line $y = 1$ to give the image B

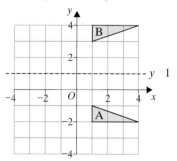

Regular polygon – a polygon with all sides and all angles equal

Regular tetrahedron – a triangular pyramid with equilateral triangles as its sides

Relative frequency – see experimental probability

Rhombus – a quadrilateral with four equal sides and opposite sides parallel

Right-angled triangle – a triangle with one angle of 90°

Sample space diagram – see two-way table

Scalar – a quantity (size) that has magnitude but not direction, for example, the numbers 2, 3, 4, ...

Scale factor – the ratio of corresponding sides usually expressed numerically so that:

$$\text{Scale factor} = \frac{\text{length of line on the enlargement}}{\text{length of line on the original}}$$

Sector (of a circle) – a region in a circle bounded by two radii and an arc

Segment – the region bounded by an arc and a chord

Major segment

Minor segment

Side elevation – a diagram of a 3-D solid showing the view from the side; sometimes the elevation from the side of the shape is the same as the front elevation, for example,

Front/side elevation

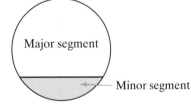

4 cm

Side

3 cm

Front

4 cm

3 cm

Usually, however, they will be different, for example,

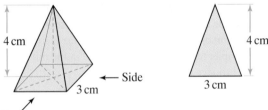

S

F

Front elevation viewed from F

Side elevation viewed from S

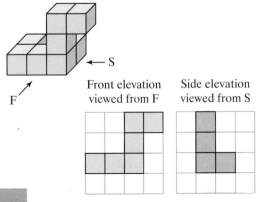

Unseen edges are shown as dotted lines, for example, the side elevation of this cylindrical container has three unseen edges. (The front elevation is the same)

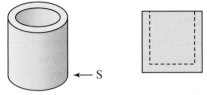

S

Similar – shapes are similar if their corresponding angles are equal *and* their corresponding sides are in the same ratio, so that one shape may be an enlargement of the other

Simplify – to make simpler by collecting like terms

Simultaneous equations – two equations that apply simultaneously to given variables; the solution to the simultaneous equations is the pair of values for the variables that satisfies both equations; the *graphical* solution to simultaneous equations is a point where the lines representing the equations *intersect*

Sine (abbreviation sin) – in a right-angled triangle, the ratio of the length of the opposite side to the length of the hypotenuse

$$\sin x = \frac{opposite}{hypotenuse}$$

Sketch – an outline of the graph with key features defined, for example, where the graph crosses the *x*-axis and *y*-axis

Solid – a three-dimensional shape

Solution – the value of the unknown in an equation, for example, the solution of the equation $3y = 6$ is $y = 2$

Solve – when you solve an equation you find the solution, in an equation or expression

Speed – the rate of change of distance with respect to time. To calculate the average speed, divide the total distance moved by the total time taken. It is usually given in metres per second (m/s) or kilometres per hour (km/h) or miles per hour (mph)

In the triangle, cover the item you want, then the rest tells you what to do

$$\text{Speed} = \frac{\text{distance}}{\text{time}}$$

$$\frac{\text{metres}}{\text{seconds}} \text{ gives m/s}$$

$$\text{Distance} = \text{speed} \times \text{time}$$

$$\text{Time} = \frac{\text{distance}}{\text{speed}}$$

Square number – a square number is the outcome when a whole number is multiplied by itself; square numbers are 1, 4, 9, 16, 25, ...

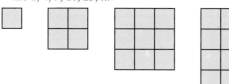

The rule for the nth term of the square numbers is n^2

Square root – the square root of a number such as 16 is a number whose outcome is 16 when multiplied by itself

Stretch – a transformation that moves lines of an object parallel to another fixed line (usually the x-axis or y-axis)

Subject of a formula – in the formula $P = 2(l + w)$, P is the subject of the formula

Substitute – find the value of an expression when the variable is given a value, for example, when $x = 4$, the expression $3x + 2 = 3 \times 4 + 2 = 14$

Subtend – when the end points of an arc are joined to a point on the circumference of a circle the angle formed is subtended by the arc

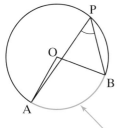

arc AB subtends $\angle$AOB at the centre

arc AB subtends $\angle$APB at the circumference

arc AB

Sum – the result of adding together two (or more) numbers, variables, terms or expressions

Supplementary – two angles are supplementary if their sum is 180°

Surd – a number containing an irrational root, for example, $\sqrt{2}$ or $3 + 2\sqrt{7}$

Tangent (abbreviation tan) – in a right-angled triangle, the ratio of the length of the opposite side to the length of the adjacent side

$$\tan x = \frac{opposite}{adjacent}$$

Tangent (to a circle) – a straight line that touches the circle at only one point

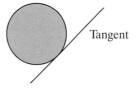

Tangent

Term – a number, variable or the product of a number and a variable(s) such as 3, x or $3x$

Theorem – a mathematical statement that can be demonstrated to be true using proof

Theoretical probability – probability based on equally likely outcomes, for example, it suggests you will get 5 heads and 5 tails if you toss a coin 10 times

Transformation – reflections, rotations and translations are transformations as they change the position but not the size of a shape

Translation – a transformation where every point moves the same distance in the same direction so that the object and the image are congruent

Shape A has been mapped onto Shape B by a translation of 3 units to the right and 2 units up

The vector for this would be $\begin{pmatrix} 3 \\ 2 \end{pmatrix}$

A translation is defined by the distance and the direction (vector)

Trapezium (pl. trapezia) – a quadrilateral with one pair of parallel sides

Trial and improvement – a method for solving algebraic equations by making an informed guess then refining this to get closer and closer to the solution

Trigonometric functions – the sine, cosine and tangent functions, $y = \sin x$, $y = \cos x$ and $y = \tan x$, are the three most common of these

The values of the sine and the cosine functions repeat every 360° and the values of the tangent function repeat every 180°, so the functions are described as **periodical**

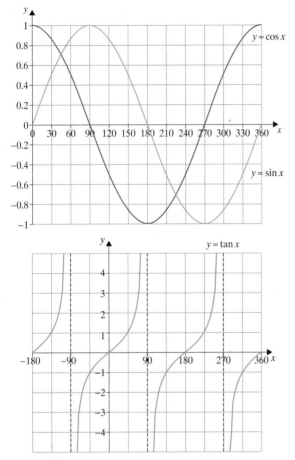

Trigonometry – the branch of mathematics that deals with the relationship between the lengths of sides and sizes of angles in triangles

Two-way table or **sample space diagram** – table used to show all the possible outcomes of an experiment, for example, all the outcomes of tossing a coin and throwing a dice

		Dice					
		1	2	3	4	5	6
Coin	Head	H1	H2	H3	H4	H5	H6
	Tail	T1	T2	T3	T4	T5	T6

Unit – a standard used in measuring, for example, a metre is a unit of length

Unknown – the letter in an equation such as x or y

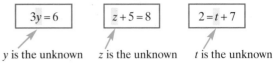

y is the unknown z is the unknown t is the unknown

Unlikely – an outcome with a probability less than $\frac{1}{2}$, for example, snow falling in August in the UK

Upper bound – this is the maximum possible value of a measurement, for example, if a length is measured as 37 cm correct to the nearest centimetre, the upper bound of the length is 37.5 cm

Variable – a symbol representing a quantity that can take different values such as x, y or z

Vector – a quantity with direction and magnitude (size)

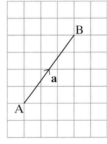

In this diagram, the arrow represents the direction and the length of the line represents the magnitude

In print, this vector can be written as **AB** or **a**

In handwriting, this vector is usually written as $\overrightarrow{AB}$ or $\underline{a}$

The vector can also be described as a column

vector $\begin{pmatrix} 3 \\ 4 \end{pmatrix}$

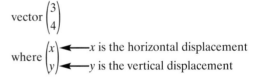

where $\begin{pmatrix} x \\ y \end{pmatrix}$ ← x is the horizontal displacement, y is the vertical displacement

Velocity – the rate of change of displacement with respect to time

Vertex (pl. **vertices**) – the point where two or more edges meet